CAMBRIDGE TRACTS IN MATHEMATICS

General Editors

230 **Fractional Sobolev Spaces and Inequalities**

CAMBRIDGE TRACTS IN MATHEMATICS

GENERAL EDITORS

J. BERTOIN, B. BOLLOBÁS, W. FULTON, B. KRA, I. MOERDIJK,
C. PRAEGER, P. SARNAK, B. SIMON, B. TOTARO

A complete list of books in the series can be found at www.cambridge.org/mathematics.
Recent titles include the following:

Fractional Sobolev Spaces and Inequalities

D. E. EDMUNDS
University of Sussex

W. D. EVANS
Cardiff University

CAMBRIDGE
UNIVERSITY PRESS

CAMBRIDGE
UNIVERSITY PRESS

University Printing House, Cambridge CB2 8BS, United Kingdom

One Liberty Plaza, 20th Floor, New York, NY 10006, USA

477 Williamstown Road, Port Melbourne, VIC 3207, Australia

314–321, 3rd Floor, Plot 3, Splendor Forum, Jasola District Centre,
New Delhi – 110025, India

103 Penang Road, #05–06/07, Visioncrest Commercial, Singapore 238467

Cambridge University Press is part of the University of Cambridge.

It furthers the University's mission by disseminating knowledge in the pursuit of
education, learning, and research at the highest international levels of excellence.

www.cambridge.org
Information on this title: www.cambridge.org/9781009254632
DOI: 10.1017/9781009254625

First published 2023

A catalogue record for this publication is available from the British Library.

ISBN 978-1-009-25463-2 Hardback

Contents

Preface

Two of the most important and useful inequalities in the theory of differential equations are the Hardy inequality

$$\left|\frac{p-n}{p}\right|^p \int_{\mathbb{R}^n} \frac{|u(x)|^p}{|x|^p}\, dx \le \int_{\mathbb{R}^n} |\nabla u(x)|^p dx\, (n \in \mathbb{N}, 1 < p < \infty)$$

and that of Rellich:

$$\left(\frac{n(p-1)(n-2p)}{p^2}\right)^p \int_{\mathbb{R}^n} \frac{|u(x)|^p}{|x|^{2p}}\, dx \le \int_{\mathbb{R}^n} |\Delta u(x)|^p\, dx (n \in \mathbb{N}, n > 2p),$$

each inequality holding for an appropriate class of functions u. Details of the long history and wide applications of the Hardy inequality are given in [15] and [110] (see also [108]); note that the special case $p = 2$ of it can be regarded as a mathematical representation of Heisenberg's uncertainty principle. As for the Rellich inequality, we refer to [15] for background information and observe that the case $p = 2$ is again distinguished in that it has implications for the self-adjointness problem for Schrödinger operators with singular potentials.

Motivated by the demands of various applications (see, for example, Chapter 1 of [40]), fractional versions of these results have been obtained in recent times. For example, when $s \in (0, 1)$, a fractional analogue of the Hardy inequality takes the form

$$c(n, s, p) \int_{\mathbb{R}^n} \frac{|u(x)|^p}{|x|^{ps}}\, dx \le \int_{\mathbb{R}^n} \int_{\mathbb{R}^n} \frac{|u(x) - u(y)|^p}{|x - y|^{n+sp}}\, dx\, dy$$

for all u in a class of functions depending on whether $ps - n$ is positive or negative. There are important applications for versions of the classical and fractional inequalities in which integration occurs over an open subset Ω of $\mathbb{R}^n$; these present interesting and challenging problems involving the geometry of Ω. The Rellich inequality has also enjoyed substantial development in the last few years.

The natural setting for these later inequalities is that of fractional Sobolev spaces which, after their introduction in the 1950s by Aronszajn, Gagliardo and Slobodeckij, have found applications in a vast number of questions involving differential equations and nonlocal effects: see, for example, the references given in [142]. Details of the historical background are given in [171]. Our objective in this book is to present an introduction to such spaces and to go on to establish inequalities such as those mentioned above.

Chapter 1 is devoted to topics that are mainly quite familiar, and are given here for the convenience of the reader and also to establish some standard notation. There follows a brief account of classical Sobolev spaces, including some of the basic embedding theorems. In Chapter 3 fractional Sobolev spaces on an open subset Ω of $\mathbb{R}^n$ are introduced: if $s \in (0, 1)$ and $p \in [1, \infty)$, such a space is

$$
W_p^s(\Omega) := \left\{ u \in L_p(\Omega) \colon (x, y) \longmapsto \frac{|u(x) - u(y)|}{|x - y|^{\frac{n}{p}+s}} \in L_p(\Omega \times \Omega) \right\},
$$

and is endowed with the norm

$$
\|u\|_{s,p,\Omega} := \left(\int_\Omega |u|^p \, dx + \int_\Omega \int_\Omega \frac{|u(x) - u(y)|^p}{|x - y|^{n+sp}} \, dx \, dy \right)^{1/p}.
$$

The term

$$
[u]_{s,p,\Omega} := \left(\int_\Omega \int_\Omega \frac{|u(x) - u(y)|^p}{|x - y|^{n+sp}} \, dx \, dy \right)^{1/p}
$$

featured here is the so-called Gagliardo seminorm; it plays the role occupied by the L_p norm of the gradient in the classical first-order Sobolev space. Note that the condition $s \in (0, 1)$ is essential if triviality is to be avoided, for when $s \geq 1$ and Ω are connected, then by Proposition 2 of [33] the only measurable functions such that

$$
\int_\Omega \int_\Omega \frac{|u(x) - u(y)|^p}{|x - y|^{n+sp}} \, dx \, dy < \infty
$$

are constants. When the underlying space Ω is bounded and has sufficiently smooth boundary these fractional spaces can be identified with certain Besov spaces. This enables such matters as embedding theorems to be taken over from the theory of Besov spaces. However, the loss of control of constants (in embedding inequalities, for example) and the conditions imposed on the boundary make it desirable to supplement the general impression given by the Besov space approach by arguments based on the concrete definition of the fractional spaces. These spaces not only have similarities to those of classical Sobolev type but also exhibit differences: for example, the Poincaré inequality holds for all bounded open subsets of $\mathbb{R}^n$ in the fractional case but not in the classical situation. Details of the behaviour of the Gagliardo seminorm as $s \to 1-$ and $s \to 0+$ (see [24], [33], [34] and [134]) are presented together with an indication of an alternative approach given by Milman et al. ([136], [107]) using interpolation theory. For additional illustration of the role that interpolation can play in fractional spaces see [31]. The chapter concludes with a brief discussion of fractional powers of the Laplacian (in connection with which we mention [112]) and associated eigenvalue problems. Chapter 4 provides a brief look at eigenvalue problems set in fractional spaces: we describe some of the most fundamental results relating to the fractional Laplacian and fractional p-Laplacian in the hope that those unfamiliar with this material will find it as fascinating as we do.

Chapter 5 presents a survey of results on Hardy inequalities in the context of classical Sobolev spaces typified by the inequality

$$\int_\Omega |\nabla u(x)|^p dx \geq C(p, \Omega) \int_\Omega \frac{|u(x)|^p}{\delta(x)^p} \, dx, \ u \in C_0^1(\Omega),$$

where Ω is an open set in $\mathbb{R}^n$, $n \geq 1$, $p \in (0, \infty)$, $\delta(x) = \inf\{|x - y| : y \in \mathbb{R}^n \setminus \Omega\}$, and $C(p, \Omega)$ is a positive constant which depends on p and Ω but not on u. Properties of the distance function δ depend on the geometry of Ω and its boundary, and these are important features of the inequality. Hardy inequalities have always attracted a good deal of interest, but the volume of high-quality work in this area does seem to have grown dramatically in this century. The literature is now so enormous as to make the selection of results to include in this book difficult; the choice is inevitably personal and some significant results are bound to have been omitted. Although it is not possible in these pages to give proofs of all the results that we do mention, precise references are provided when proofs are not. We also present the inequality of Laptev and Weidl on $\mathbb{R}^2 \setminus \{0\}$ in which the gradient ∇ of Hardy's inequality is replaced by the magnetic gradient $\nabla + \nu A$, where A is a magnetic potential of Aharonov–Bohm type. Discrete versions of the resulting inequality are also discussed. The Hardy theme is continued in Chapter 6 in the fractional setting.

Finally, in Chapter 7 the focus is on the Rellich inequality. To set the scene, results obtained in classical Sobolev spaces are surveyed and then some recent developments of a fractional nature are presented.

Chapters are divided into sections and sections are sometimes divided into subsections. Theorems, Corollaries, Lemmas, Propositions, Remarks and equations are numbered consecutively. At the end of the book there are author, subject and notation indexes.

Basic Notation

$\mathbb{N}$: set of all natural numbers.

$\mathbb{N}_0 = \mathbb{N} \cup \{0\}$.

$\mathbb{R}$: set of all real numbers.

$\mathbb{R}^n$: n-dimensional Euclidean space.

$\mathbb{R}^n_+ : \{(x_i) \in \mathbb{R}^n : x_n > 0\}$.

ω_n: volume of unit ball in $\mathbb{R}^n$.

Ω: open subset of $\mathbb{R}^n$ with closure $\overline{\Omega}$ and boundary $\partial\Omega$.

$B(X, Y)$: space of all bounded linear maps from a Banach space X to another such space Y.

$K(X, Y)$: subspace of $B(X, Y)$ consisting of all compact linear maps from X to Y.

Embedding: a bounded linear injective map of a Banach space X to another such space Y.

$X \hookrightarrow Y$: the space X is embedded in Y.

$X \hookrightarrow\hookrightarrow Y$: the space X is compactly embedded in Y.

$L_p(\Omega)$: the Lebesgue space of all scalar-valued functions f on Ω such that $\int_\Omega |f|^p \, dx < \infty \, (1 \leq p < \infty)$.

$L_\infty(\Omega)$: the Lebesgue space of all scalar-valued functions f on Ω such that ess $\sup_\Omega |f(x)| < \infty$.

F: Fourier transform given by $F(f)(\xi) = (2\pi)^{-n/2} \int_{\mathbb{R}^n} f(x) e^{-ix\cdot\xi} \, dx$.

1

Preliminaries

1.1 Integration

We presuppose basic knowledge of the theory of Lebesgue integration on measurable subsets of $\mathbf{R}^n$. The few results listed below are often used subsequently and are given simply for the convenience of the reader. Proofs may be found in [54] and [157], for example.

By Ω we shall usually mean a measurable (often open) subset of $\mathbf{R}^n$; its Lebesgue n-measure will be denoted by $|\Omega|_n$, or even by $|\Omega|$ if no ambiguity is possible. All functions mentioned in this subsection are assumed to be extended real-valued; given any such function f on Ω, we set

$$f^+ = \max\{f, 0\}, f^- = -\min\{f, 0\}.$$

A measurable function f on Ω is said to be integrable over Ω if both $\int_\Omega f^+(x)\,dx$ and $\int_\Omega f^-(x)\,dx$ are finite.

Theorem 1.1 (The monotone convergence theorem) *Let $\{f_k\}_{k\in\mathbb{N}}$ be a non-decreasing sequence of measurable functions on Ω such that for some $k \in \mathbb{N}$, $\int_\Omega f_k^-(x)\,dx < \infty$. Then*

$$\lim_{l\to\infty} \int_\Omega f_l(x)\,dx = \int_\Omega \lim_{l\to\infty} f_l(x)\,dx.$$

Theorem 1.2 (Fatou's lemma) *Let $\{f_k\}_{k\in\mathbb{N}}$ be a sequence of non-negative measurable functions on Ω. Then*

$$\int_\Omega \liminf_{k\to\infty} f_k(x)\,dx \leq \liminf_{k\to\infty} \int_\Omega f_k(x)\,dx.$$

Theorem 1.3 (Lebesgue's dominated convergence theorem) *Let $\{f_k\}_{k\in\mathbb{N}}$ be a sequence of measurable functions on Ω such that for almost all $x \in \Omega$, $\lim_{k\to\infty} f_k(x) = f(x)$. Moreover, suppose that there is a function g, integrable over Ω, such that*

$$|f_k(x)| \leq g(x) \text{ for all } k \in \mathbb{N} \text{ and almost all } x \in \Omega.$$

Then f and each f_k are integrable over Ω and

$$\int_\Omega f(x)\, dx = \lim_{k\to\infty} \int_\Omega f_k(x)\, dx.$$

Theorem 1.4 (Fubini's theorem)   *For $i = 1, 2$, let Ω_i be a measurable subset of $\mathbb{R}^{n_i}$; put $\Omega = \Omega_1 \times \Omega_2$ and suppose that $f\colon \Omega \to \mathbb{R}$ is such that $\int_\Omega f(x, y)\, dx\, dy$ is finite. Then $\int_{\Omega_1} f(x, y)\, dx$ exists for almost all $y \in \Omega_2$, $\int_{\Omega_2} f(x, y)\, dy$ exists for almost all $x \in \Omega_1$, and*

$$\int_\Omega f(x, y)\, dx\, dy = \int_{\Omega_1}\left(\int_{\Omega_2} f(x, y)\, dy\right) dx = \int_{\Omega_2}\left(\int_{\Omega_1} f(x, y)\, dx\right) dy.$$

To apply this theorem we need to know that the function f is integrable over Ω. This difficulty is overcome by Tonelli's theorem, which leads to the conclusion that if one of the iterated integrals

$$\int_{\Omega_1}\left(\int_{\Omega_2} |f(x, y)|\, dy\right) dx, \quad \int_{\Omega_2}\left(\int_{\Omega_1} |f(x, y)|\, dx\right) dy$$

is finite, then f is integrable over Ω and the conclusion of Theorem 1.4 holds. For details of this, see [54], p. 194 and [164], pp. 353–354.

The next result gives connections between various types of convergence of functions.

Theorem 1.5   *Let $p \in [1, \infty)$, let Ω be a measurable subset of $\mathbb{R}^n$ and suppose that $f, f_k\ (k \in \mathbb{N})$ are functions on Ω such that*

$$\int_\Omega |f(x)|^p\, dx < \infty, \quad \int_\Omega |f_k(x)|^p\, dx < \infty \ (k \in \mathbb{N})$$

and

$$\int_\Omega |f(x) - f_k(x)|^p\, dx \to 0 \text{ as } k \to \infty.$$

Then:

 (i) *There is a subsequence of $\{f_k\}$ that converges pointwise a.e. to f.*
 (ii) *The sequence $\{f_k\}$ **converges in measure** to f: that is, given any $\varepsilon > 0$,*

$$\lim_{k\to\infty} |\{x \in \Omega\colon |f_k(x) - f(x)| > \varepsilon\}| = 0.$$

We shall occasionally need to deal with integration over σ-finite measure spaces: details of this, which follow similar lines to that just detailed, may be found in Chapter 1 of [146].

1.2 Banach Spaces

It is assumed that the reader is familiar with the fundamental concepts concerning normed linear spaces: our purpose here is to place on record the notation and some basic facts. More details and proofs of the various assertions may be found in standard texts on functional analysis, such as [54].

Given a normed linear space X over the real or complex field, its norm will be denoted by $\|\cdot|X\|$ or $\|\cdot\|_X$, depending on the size of the expression X; if there is no ambiguity, we shall simply write $\|\cdot\|$. The closed ball in X with centre x and radius r is represented by $B(x, r)$, abbreviated to B_X if $x = 0$ and $r = 1$; by S_X will be meant the unit sphere $\{x \in X: \|x\| = 1\}$. A *Banach space X* is a normed linear space that is complete in the sense that every Cauchy sequence in X converges to a point in X. Let X, Y be Banach spaces over the same field of scalars and let $T: X \to Y$ be linear. Then T is continuous if and only if

$$\|T\| := \sup \{\|Tx\|_Y : \|x\|_X \le 1\} < \infty;$$

$B(X, Y)$ stands for the set of all continuous linear maps from X to Y, abbreviated to $B(X)$ if $X = Y$. The map $T \longmapsto \|T\|$ is a norm on this space endowed with which $B(X, Y)$ is a Banach space. The *dual X^** of X is the space $B(X, \Phi)$, where Φ is the underlying scalar field. Given $x \in X$ and $x^* \in X^*$, we shall often denote $x^*(x)$ by $\langle x, x^* \rangle_X$, or even $\langle x, x^* \rangle$ if the context is clear. A sequence $\{x_j\}_{j \in \mathbb{N}}$ in X converges *strongly* to $x \in X$, written $x_j \to x$, if and only if $\|x - x_j\| \to 0$; it converges *weakly* to x, written $x_j \rightharpoonup x$, if and only if $\langle x_j - x, x^* \rangle \to 0$ for all $x^* \in X^*$. The *adjoint* of a map $T \in B(X, Y)$ is the map $T^*: Y^* \to X^*$ defined by

$$\langle x, T^* y^* \rangle_X = \langle Tx, y^* \rangle_Y \text{ for all } x \in X \text{ and all } y^* \in Y^*.$$

It emerges that $T^* \in B(Y^*, X^*)$ and $\|T^*\| = \|T\|$. A linear map $T: X \to Y$ is said to be *compact* if, for every bounded set $B \subset X$, the closure $\overline{T(B)}$ is compact in Y; equivalently, given any bounded sequence $\{x_n\}$ in X, $\{Tx_n\}$ has a subsequence that converges in Y. A compact linear map is necessarily bounded; the family $K(X, Y)$ of all compact linear maps from X to Y is closed in $B(X, Y)$.

A map $T \in B(X, Y)$ is said to be *strictly singular* if there is no infinite dimensional closed subspace Z of X such that the restriction $T|_Z$ of T to Z is an isomorphism of Z onto $T(Z)$. Equivalently, for each infinite-dimensional closed subspace Z of X,

$$\inf \{\|Tx\|_Y : \|x\|_X = 1, x \in Z\} = 0.$$

If instead T has the property that given any $\varepsilon > 0$ there exists $N(\varepsilon) \in \mathbb{N}$ such that if E is a subspace of X with $\dim E \ge N(\varepsilon)$, then there exists $x \in E$, with

$\|x\|_X = 1$, such that $\|Tx\|_Y \leq \varepsilon$, then T is said to be *finitely strictly singular*. This second definition can be expressed in terms of the Bernstein numbers $b_k(T)$ of T. We recall that these are given, for each $k \in \mathbb{N}$, by

$$b_k(T) = \sup_{E \subset X, \dim E = k} \inf_{x \in E, \|x\|_X = 1} \|Tx\|_X.$$

Then T is finitely strictly singular if and only if

$$b_k(T) \to 0 \text{ as } k \to \infty.$$

The relations between these notions and that of compactness of T are illustrated by the following diagram:

$$T \text{ compact} \implies T \text{ finitely strictly singular} \implies T \text{ strictly singular}$$

and each reverse implication is false in general. For further details and general background information concerning these matters, together with particular examples, we refer to [1], [117], [118], [119] and [147].

We write $X \hookrightarrow Y$ to signify that X can be identified with a subset of Y and that the natural embedding map from X to Y is continuous; if this map is compact we write $X \hookrightarrow\hookrightarrow Y$. The dual X^* of a Banach space X is also a Banach space, the dual of which is denoted by X^{**}. There is a natural mapping $\kappa: X \to X^{**}$ defined by

$$\langle x^*, \kappa x \rangle_{X^*} = \langle x, x^* \rangle_X \text{ for all } x \in X \text{ and all } x^* \in X^*;$$

κ is an isometric isomorphism of X onto $\kappa(X)$. If $\kappa(X) = X^{**}$ the space X is said to be *reflexive*. An important property of any reflexive space X is that every bounded sequence in X has a subsequence that is weakly convergent to some point of X.

The *modulus of convexity* of a Banach space X (with $\dim X \geq 2$) is the map $\delta_X: (0, 2] \to [0, 1]$ defined by

$$\delta_X(\varepsilon) = \inf\left\{1 - \frac{1}{2}\|x + y\| : x, y \in B_X, \|x - y\| \geq \varepsilon\right\};$$

the space X is said to be *uniformly convex* if $\delta_X(\varepsilon) > 0$ for all $\varepsilon \in (0, 2]$; every uniformly convex space is reflexive. Morover, if $\{x_j\}_{j \in \mathbb{N}}$ is a sequence in a uniformly convex space X such that $x_j \rightharpoonup x \in X$ and $\|x_j\| \to \|x\|$, then $x_j \to x$. For details of the companion notion of uniform smoothness see [61], Chapter 1.

A Banach space X is said to have the *approximation property* (AP) if, given any compact subset K of X and any $\varepsilon > 0$, there exists $T \in B(X)$ with finite rank such that $\|Tx - x\| < \varepsilon$ for all $x \in K$. Every Banach space X with a basis has the AP: we recall that a sequence $\{x_n\}_{n \in \mathbb{N}}$ of elements of X is a (Schauder) *basis* of X if, given any $x \in X$, there is a unique sequence $\{a_n\}_{n \in \mathbb{N}}$ of scalars such that $x = \sum_{n=1}^{\infty} a_n x_n$. We refer to Chapter 1 of [61] and the references given there for proofs of these assertions, together with that made below, and further background.

A map $\mu \colon [0, \infty) \to [0, \infty)$ that is continuous, strictly increasing and has the properties that $\mu(0) = 0$ and $\lim_{t \to \infty} \mu(t) = \infty$ is called a *gauge function*. Given a gauge function μ and a Banach space X with uniformly convex dual X^*, there is a map $\kappa \colon X \to X^*$ such that for each $x \in X$, $\kappa x = x^*$, where

$$\langle x, x^* \rangle = \|x\| \, \|x^*\| \quad \text{and} \quad \|x^*\| = \mu\left(\|x\|\right).$$

That this map is well defined is a consequence of the uniform convexity of X^*: indeed this is the case under weaker assumptions on X^*. The map κ is called the *duality map* on X (with gauge function μ) and is continuous on X: $\kappa x_k \to \kappa x$ in X^* whenever $x_k \to x$ in X. For proofs of these assertions and further details we refer to [61], Chapter 1.

A particularly important class of Banach spaces is that of Hilbert spaces, which we now briefly recall. An *inner product* on a linear space X over a scalar field Φ is a map $(\cdot, \cdot) \colon X \times X \to \Phi$ such that

(i) $(\alpha x_1 + \beta x_2, y) = \alpha \, (x_1, y) + \beta \, (x_2, y)$ for all $\alpha, \beta \in \Phi$ and all $x_1, x_2, y \in X$;
(ii) $(x, y) = \overline{(y, x)}$ for all $x, y \in X$;
(iii) $(x, x) > 0$ if $x \in X \backslash \{0\}$.

A linear space X equipped with an inner product is called an *inner product space*; the map $x \mapsto (x, x)^{1/2}$ is a norm on X; and if the resulting normed linear space is complete, it is said to be a *Hilbert space*. Every Hilbert space is uniformly convex.

To conclude this section we give various examples of Banach spaces.

(i) $\mathbb{R}^n$ and $\mathbb{C}^n$ with norm given by $\|x\| = \left(\sum_{j=1}^{n} |x_j|^2\right)^{1/2}$, $x = (x_1, ..., x_n)$;
these are Hilbert spaces, with the natural definition of the inner product.
(ii) l_p, the space of all sequences $x = \{x_j\}_{j \in \mathbb{N}}$ of scalars such that

$$\|x\|_p := \left(\sum_{j=1}^{\infty} |x_j|^p\right)^{1/p} < \infty \, (1 \le p < \infty),$$

and

$$\|x\|_\infty := \sup_{j \in \mathbb{N}} |x_j| < \infty.$$

(iii) $L_p(\Omega)$, the linear space of all (Lebesgue) measurable functions on a measurable subset Ω of $\mathbb{R}^n$, functions equal almost everywhere being identified, such that

$$\|f\|_{p,\Omega} := \left(\int_\Omega |f(x)|^p\,dx\right)^{1/p} < \infty\ (1 \le p < \infty),$$

and

$$\|f\|_{\infty,\Omega} := \operatorname*{ess\,sup}_\Omega |f(x)|.$$

When $1 < p < \infty$, the spaces l_p and $L_p(\Omega)$ are uniformly convex and have the AP; they are Hilbert spaces, with natural definitions of the inner product, if $p = 2$. The duality map κ on $L_p(\Omega)$ $(1 < p < \infty)$ with gauge function $t \longmapsto t^{p-1}$ is given by $\kappa f = |f|^{p-2}f$ $\left(f \in L_p(\Omega)\right)$.

1.3 Function Spaces

1.3.1 Spaces of Continuous Functions

Throughout, Ω will stand for a non-empty open subset of $\mathbb{R}^n$ with boundary $\partial\Omega$ and closure $\overline{\Omega}$; a *domain* is a connected open set. Points of $\mathbb{R}^n$ will be denoted by $x = (x_i) = (x_1, ..., x_n)$ and we write $|x| = \left(\sum_{i=1}^n x_i^2\right)^{1/2}$ and $(x,y) = \sum_{i=1}^n x_i y_i$; given $r > 0$, we put $B(x,r) = \{y \in \mathbb{R}^n: |x-y| < r\}$, abbreviating this to B_r if $x = 0$. If $\alpha = (\alpha_1, ..., \alpha_n) \in \mathbb{N}_0^n$, where $\mathbb{N}_0 = \mathbb{N} \cup \{0\}$, we write

$$\alpha! = \prod_{j=1}^n \alpha_j!,\ |\alpha| = \sum_{j=1}^n \alpha_j,\ x^\alpha = \prod_{j=1}^n x_j^{\alpha_j}\ (x \in \mathbb{R}^n)$$

and

$$D^\alpha := \frac{\partial^{|\alpha|}}{\partial x_1^{\alpha_1}...\partial x_n^{\alpha_n}} := \prod_{j=1}^n D_j^{\alpha_j},\ \text{where } D_j = \partial/\partial x_j;$$

it is to be understood that if some α_j is zero, then the corresponding term is to be omitted; if all α_j are zero, so that $\alpha = 0$, then $D^\alpha u = u$ for any appropriate function u.

Given any $k \in \mathbb{N}_0$, by $C^k(\Omega)$ is meant the linear space of all real- or complex-valued functions u on Ω such that for all $\alpha \in \mathbb{N}_0^n$ with $|\alpha| \le k$, the function $D^\alpha u$ exists and is continuous on Ω. The subspace of $C^k(\Omega)$ consisting of all those functions with compact support contained in Ω is denoted by $C_0^k(\Omega)$, and $C_0^\infty(\Omega) := \cap_{k=1}^\infty C_0^k(\Omega)$; recall that the *support* of a function u, supp u, is the closure of $\{x \in \Omega: u(x) \ne 0\}$. The function ϕ defined on $\mathbb{R}^n$ by

$$\phi(x) = \begin{cases} \exp\left(\frac{-1}{1-|x|^2}\right), & |x| < 1, \\ 0, & |x| \ge 1 \end{cases}$$

belongs to $C_0^\infty(\mathbb{R}^n)$, with supp $\phi = \overline{B(0,1)}$ and $\int_{\mathbb{R}^n} \phi(x)\,dx > 0$, so that $\psi := \phi/\int_{\mathbb{R}^n}\phi(x)\,dx$ has the useful properties that $\psi \in C_0^\infty(\mathbb{R}^n)$ and $\int_{\mathbb{R}^n}\psi(x)\,dx = 1$.

We define $C^k(\overline{\Omega})$ to be the linear space of all bounded functions u in $C^k(\Omega)$ such that u and all its derivatives $D^\alpha u$ with $|\alpha| \le k$ have bounded, continuous extensions to $\overline{\Omega}$: a norm $||| \cdot |||_{k,\Omega}$ is defined on this space by

$$||| u |||_{k,\Omega} := \max_{|\alpha| \le k} \sup_{x \in \Omega} |D^\alpha u(x)|,$$

and $C^k(\overline{\Omega})$ becomes a Banach space when given this norm. For a discussion of the advantages and disadvantages of this notation see [65], p. 10.

Let $k \in \mathbb{N}_0$, $\lambda \in (0, 1]$. We shall need various spaces of Hölder-continuous functions. First, $C^{0,\lambda}(\Omega)$ (often written as $C^\lambda(\Omega)$) will stand for the linear space of all continuous functions on Ω which satisfy a local Hölder condition on Ω; that is, given any compact subset K of Ω, there is a constant $C > 0$ such that

$$|u(x) - u(y)| \le C |x - y|^\lambda \text{ for all } x, y \in K.$$

We also put

$$C^{k,\lambda}(\Omega) := \left\{ u \in C^k(\Omega) : D^\alpha u \in C^\lambda(\Omega) \text{ for all } \alpha \in \mathbb{N}_0^n \text{ with } |\alpha| = k \right\}.$$

These spaces are not given norms. However,

$$C^{k,\lambda}(\overline{\Omega}) := \left\{ u \in C^k(\overline{\Omega}) : \text{given any } \alpha \in \mathbb{N}_0^n \text{ with } |\alpha| = k, \text{ there} \right.$$
$$\left. \text{exists } C > 0 \text{ such that for all } x, y \in \Omega, |D^\alpha u(x) - D^\alpha u(y)| \le C |x - y|^\lambda \right\}$$

becomes a Banach space when provided with the norm

$$||| u |||_{k,\lambda,\Omega} := ||| u |||_{k,\Omega} + |u|_{k,\lambda,\Omega},$$

where

$$|u|_{k,\lambda,\Omega} = \max_{|\alpha|=k} \sup_{x,y \in \Omega, x \ne y} |D^\alpha u(x) - D^\alpha u(y)| / |x - y|^\lambda.$$

For convenience, when $\lambda \in (0, 1)$, we write $C^\lambda(\overline{\Omega}) = C^{0,\lambda}(\overline{\Omega})$ and $||| \cdot |||_{\lambda,\Omega}$, $|\cdot|_{\lambda,\Omega}$ instead of $||| \cdot |||_{0,\lambda,\Omega}$, $|\cdot|_{0,\lambda,\Omega}$ respectively. By $C_0^{k,\lambda}(\Omega)$ will be meant the linear subspace of $C^{k,\lambda}(\Omega)$ consisting of all those functions with compact support contained in Ω. Note that if $u, v \in C^\lambda(\overline{\Omega})$, then

$$|uv|_{\lambda,\Omega} \le ||| u |||_{0,\Omega}|v|_{\lambda,\Omega} + ||| v |||_{0,\Omega}|u|_{\lambda,\Omega};$$

and that if $u \in C^{\lambda_1}(\overline{\Omega})$, $v \in C^{\lambda_2}(\overline{\Omega})$ and Ω is bounded, then $uv \in C^\gamma(\overline{\Omega})$, where $\gamma = \min(\lambda_1, \lambda_2)$, and

$$||| uv |||_{\gamma,\Omega} \le \max \left\{ 1, |\text{diam } \Omega|^{\lambda_1 + \lambda_2 - 2\gamma} \right\} ||| u |||_{\lambda_1,\Omega} ||| v |||_{\lambda_2,\Omega}.$$

Useful properties relating these spaces of functions are given in the following theorem.

Theorem 1.6 *Let* $k \in \mathbb{N}_0$, $0 < \nu < \lambda \leq 1$ *and suppose that* Ω *is an open subset of* $\mathbb{R}^n$. *Then*

$$(i) \ C^{k+1}\left(\overline{\Omega}\right) \hookrightarrow C^k\left(\overline{\Omega}\right)$$

and

$$(ii) \ C^{k,\lambda}\left(\overline{\Omega}\right) \hookrightarrow C^{k,\nu}\left(\overline{\Omega}\right) \hookrightarrow C^k\left(\overline{\Omega}\right).$$

If Ω *is bounded, both the embeddings in (ii) are compact. If* Ω *is convex, then*

$$(iii) \ C^{k+1}\left(\overline{\Omega}\right) \hookrightarrow C^{k,1}\left(\overline{\Omega}\right)$$

and

$$(iv) \ C^{k+1}\left(\overline{\Omega}\right) \hookrightarrow C^{k,\nu}\left(\overline{\Omega}\right).$$

If Ω *is bounded and convex, then the embeddings in (i) and (iv) are compact.*

Finally we turn to conditions useful in the extension of functions. Let Ω be an open subset of $\mathbb{R}^n$ ($n \geq 2$) with non-empty boundary $\partial\Omega$, let $k \in \mathbb{N}_0$ and suppose that $\gamma \in [0,1]$. Given $x_0 \in \partial\Omega$, $r > 0$, $\beta > 0$, local Cartesian co-ordinates $y = (y_1, ..., y_n) = (y', y_n)$ (where $y' = (y_1, ..., y_{n-1})$), with $y = 0$ at $x = x_0$, and a real continuous function $h: y' \longmapsto h(y')$ ($|y'| < r$), we define a neighbourhood $U_{r,\beta,h}(x_0)$ of x_0 (an open subset of $\mathbb{R}^n$ containing x_0) by

$$U = U_{r,\beta,h}(x_0) = \left\{y \in \mathbb{R}^n : h(y') - \beta < y_n < h(y') + \beta, |y'| < r\right\}.$$

Then Ω is said to have boundary $\partial\Omega$ of class $C^{k,\gamma}$ if for each $x_0 \in \partial\Omega$ there are a local co-ordinate system, positive constants r and β and a function $h \in C^{k,\gamma}\left(B_r'\right)$ (where $B_r' = \left\{y' \in \mathbb{R}^{n-1} : |y'| < r\right\}$) such that

$$U_{r,\beta,h}(x_0) \cap \partial\Omega = \left\{y \in \mathbb{R}^n : y_n = h(y'), |y'| < r\right\}$$

and

$$U_{r,\beta,h}(x_0) \cap \Omega = \left\{y \in \mathbb{R}^n : h(y') - \beta < y_n < h(y'), |y'| < r\right\}.$$

In general, the constants r, β and the function h depend on x_0. However, if in addition Ω is bounded, there are points $x_1, ..., x_m \in \partial\Omega$, positive numbers r and β (independent of the x_j) and functions $h_1, ..., h_m$ such that the neighbourhoods $U_j = U_{r,\beta,h_j}(x_j)$ ($j = 1, ..., m$) cover $\partial\Omega$. When $\gamma = 0$ we simply write $\partial\Omega \in C^k$ (or $\partial\Omega \in C$ if $k = 0$). If $\partial\Omega \in C^{0,1}$ we shall say that the boundary is of Lipschitz class: if Ω is convex its boundary is of this class.

1.3.2 Morrey and Campanato Spaces

These are defined by means of some kind of mean oscillation property imposed on their elements. All we need in this book is that certain Campanato spaces are isomorphic to spaces of Hölder-continuous functions, but their great importance in the past coupled with considerable current research activity lead us to give some basic definitions and results, together with references in which more details are provided.

Throughout this subsection we shall suppose that Ω is a bounded open subset of $\mathbb{R}^n$ with the property that there exists $A > 0$ such that

$$|B(x, r) \cap \Omega| \geq Ar^n \text{ for all } x \in \Omega \text{ and all } r \leq \text{diam } \Omega. \quad (1.3.1)$$

This condition means that $\partial\Omega$ cannot have sharp outward cusps; Lipschitz boundaries are allowed.

Definition 1.7 Let $p \in [1, \infty)$ and $\lambda \geq 0$. The Morrey space $M^{p,\lambda}(\Omega)$ is the space of all $u \in L_p(\Omega)$ such that

$$\left\| u | M^{p,\lambda}(\Omega) \right\|^p := \sup_{x_0 \in \Omega, 0 < r < \text{diam } \Omega} r^{-\lambda} \int_{\Omega \cap B(x_0, r)} |u(x)|^p \, dx < \infty.$$

When endowed with the norm $\left\| \cdot | M^{p,\lambda}(\Omega) \right\|$ it becomes a Banach space. The Campanato space $\mathcal{L}^{p,\lambda}(\Omega)$ is the space of all $u \in L_p(\Omega)$ such that

$$\{u\}_{p,\lambda}^p := \sup_{x_0 \in \Omega, 0 < r < \text{diam } \Omega} r^{-\lambda} \int_{\Omega \cap B(x_0, r)} \left| u(x) - u_{x_0, r} \right|^p \, dx < \infty,$$

where

$$u_{x_0, r} = |\Omega \cap B(x_0, r)|^{-1} \int_{\Omega \cap B(x_0, r)} u(x) \, dx.$$

Furnished with the norm

$$\left\| u | \mathcal{L}^{p,\lambda}(\Omega) \right\| := \|u\|_{p,\Omega} + \{u\}_{p,\lambda},$$

it is a Banach space. The space $\mathcal{L}^{p,\lambda}(\mathbb{R}^n)$ is defined analogously.

By way of background we list some of the main properties of these spaces.

(i) For all $p \in (1, \infty)$, both $M^{p,0}(\Omega)$ and $\mathcal{L}^{p,0}(\Omega)$ are isomorphic to $L_p(\Omega)$; $M^{p,n}(\Omega)$ is isomorphic to $L_\infty(\Omega)$.

(ii) If $1 \leq p \leq q < \infty$ and λ, ν are non-negative numbers such that $(\lambda - n)/p \leq (\nu - n)/q$, then

$$M^{q,\nu}(\Omega) \hookrightarrow M^{p,\lambda}(\Omega) \text{ and } \mathcal{L}^{q,\nu}(\Omega) \hookrightarrow \mathcal{L}^{p,\lambda}(\Omega).$$

(iii) Suppose that $p \in [1, \infty)$. Then

$$\mathcal{L}^{p,\lambda}(\Omega) \text{ is isomorphic to } M^{p,\lambda}(\Omega) \text{ if } \lambda \in [0, n),$$

$$M^{p,\lambda}(\Omega) = \{0\} \text{ if } \lambda > n,$$

$$\mathcal{L}^{p,\lambda}(\Omega) \text{ is isomorphic to } C^{(\lambda-n)/p}(\overline{\Omega}) \text{ if } \lambda \in (n, n+p],$$

and

$$\mathcal{L}^{p,\lambda}(\mathbb{R}^n) \text{ is isomorphic to } C^{(\lambda-n)/p}(\mathbb{R}^n) \text{ if } \lambda \in (n, n+p].$$

For proofs of these assertions and further details, we refer to [88], [106] (especially for the claim concerning $\mathcal{L}^{p,\lambda}(\mathbb{R}^n)$), [146] and [152].

1.3.3 Banach Function Spaces

To explain what these are we begin with the notion of the non-increasing re-arrangement of a measurable function and refer to [21], [60] or [146] for further details and proofs. Let (R, μ) be a σ-finite measure space and set

$$\mathcal{M}(R, \mu) = \left\{ f : f \text{ is a measurable function on } R \text{ with values in } [-\infty, \infty] \right\},$$

$$\mathcal{M}_0(R, \mu) = \left\{ f \in \mathcal{M}(R, \mu) : f \text{ is finite } \mu - \text{a.e. on } R \right\}$$

and

$$\mathcal{M}_+(R, \mu) = \{ f \in \mathcal{M}_0(R, \mu) : f \geq 0 \}.$$

When R is a Lebesgue-measurable subset Ω of $\mathbb{R}^n$ and μ is a Lebesgue n-measure these objects are denoted by $\mathcal{M}(\Omega)$, etc. The *non-increasing re-arrangement* $f^* : [0, \infty) \to [0, \infty]$ of a function $f \in \mathcal{M}(R, \mu)$ is defined by

$$f^*(t) = \inf \{ \lambda \in (0, \infty) : \mu(\{s \in R : |f(s)| > \lambda\}) \leq t \}, t \in [0, \infty).$$

The *maximal non-increasing rearrangement* $f^{**} : (0, \infty) \to [0, \infty]$ of a function $f \in \mathcal{M}(R, \mu)$ is given by

$$f^{**}(t) = t^{-1} \int_0^t f^*(s)ds, t \in (0, \infty).$$

If $|f| \leq |g|$ μ-a.e. in R, then $f^* \leq g^*$; however, the map $f \longmapsto f^*$ does not preserve sums or products of functions and is not subadditive. By way of compensation it turns out (see Chapter 2, (3.10) of [21]) that for all $t \in (0, \infty)$ and all $f, g \in \mathcal{M}_0(R, \mu)$,

$$\int_0^t (f+g)^*(s)\, ds \leq \int_0^t f^*(s)\, ds + \int_0^t g^*(s)\, ds,$$

so that

$$(f+g)^{**} \le f^{**} + g^{**}.$$

Moreover, the *Hardy lemma* (see Chapter 2, Proposition 3.6 of [21]) asserts that if f, g are non-negative measurable functions on $(0, \infty)$ such that

$$\int_0^t f(s)\, ds \le \int_0^t g(s)\, ds$$

for all $t \in (0, \infty)$ and $h\colon (0, \infty) \to [0, \infty)$ is non-increasing, then

$$\int_0^\infty f(s)h(s)\, ds \le \int_0^\infty g(s)h(s)\, ds.$$

The *Hardy–Littlewood inequality* (see Chapter 2, Theorem 2.2 of [21]) states that for all $f, g \in \mathcal{M}_0\,(R, \mu)$,

$$\int_R |fg|\, d\mu \le \int_0^\infty f^*(t)g^*(t)\, dt.$$

If (R, μ) and (S, ν) are σ-finite measure spaces, functions $f \in \mathcal{M}_0\,(R, \mu)$ and $g \in \mathcal{M}_0\,(S, \nu)$ are said to be *equimeasurable,* and we write $f \sim g$ if $f^* = g^*$ on $(0, \infty)$.

After these preliminaries we introduce the notion of a *Banach function norm,* by which is meant a functional $\rho\colon \mathcal{M}_0\,(R, \mu) \to [0, \infty]$ such that, for all f, g and $\{f_j\}_{j \in \mathbb{N}}$ and all $\lambda \ge 0$, the following conditions are satisfied:

(P1) $\rho(f) = 0$ if and only if $f = 0$; $\rho(\lambda f) = \lambda\rho(f)$; $\rho(f+g) \le \rho(f) + \rho(g)$ (the *norm axiom*);

(P2) $f \le g$ μ–a.e. implies $\rho(f) \le \rho(g)$ (the *lattice axiom*);

(P3) $f_j \uparrow f$ μ–a.e. implies $\rho\,(f_j) \uparrow \rho(f)$ (the *Fatou axiom*);

(P4) $\rho\,(\chi_E) < \infty$ for every $E \subset R$ with finite measure (the *non-triviality axiom*);

(P5) if $E \subset R$ with $\mu(E) < \infty$, then there is a constant $C_E \in (0, \infty)$, depending only on E and ρ, such that for all $f \in \mathcal{M}_+\,(R, \mu)$,

$$\int_E f d\mu \le C_E\rho(f)$$

(the *local embedding in L_1*);
if, in addition, ρ satisfies

(P6) $\rho(f) = \rho(g)$ whenever $f^* = g^*$ ($f,\ g \in \mathcal{M}_+(R, \mu)$) (the *rearrangement-invariance axiom*),
then we say that ρ is a *rearrangement-invariant (r.i.) norm.*

If ρ satifies conditions (P1)–(P5), the space

$$X = X(\rho) := \{f \in \mathcal{M}\,(R, \mu) : \rho\,(|f|) < \infty\}$$

is said to be a *Banach function space.* With the natural linear space operations
it is easy to check that X is a linear space and $\|\cdot\|_X$, where

$$\|f\|_X := \rho(|f|),$$

is a norm on it that makes it into a Banach space. If this Banach space X also
satisfies condition (P6), it is called a *rearrangement-invariant space,* written
r.i. Note that $\|f\|_X$ is defined for every $f \in \mathcal{M}(R, \mu)$ and that $\|f\|_X < \infty$ if and
only if $f \in X$.

With any r.i. function norm ρ is associated another functional, ρ', defined for
all $g \in \mathcal{M}_+(R, \mu)$ by

$$\rho'(g) = \sup\left\{\int_R fg\,d\mu : f \in \mathcal{M}_+(R, \mu), \rho(f) \le 1\right\}.$$

This functional is also an r.i. norm and is called the *associate norm* of ρ. For
every r.i. norm ρ and every $f \in \mathcal{M}_+(R, \mu)$, it turns out that (see Chapter 1,
Theorem 2.9 of [21])

$$\rho(f) = \sup\left\{\int_R fg\,d\mu : g \in \mathcal{M}_+(R, \mu), \rho(g) \le 1\right\}.$$

If ρ is an r.i. norm, $X = X(\rho)$ is the r.i. space determined by ρ and ρ' is the
associate norm of ρ, then the function space $X(\rho')$ determined by ρ' is called
the *associate space* of X and is denoted by X'. It emerges that $\left(X'\right)' = X$; and
the Hölder inequality

$$\int_R fg\,d\mu \le \|f\|_X \|g\|_{X'}$$

holds for all $f, g \in \mathcal{M}(R, \mu)$.

From the Hardy lemma follows (see Chapter 2, Theorem 4.6 of [21]) the
Hardy–Littlewood principle, which asserts that if functions f and g satisfy the
Hardy–Littlewood–Pólya relation, defined by

$$\int_0^t f^*(s)\,ds \le \int_0^t g^*(s)\,ds, \ t \in (0, \infty),$$

sometimes denoted by $f \prec g$ in the literature, then $\|f\|_X \le \|g\|_X$ if the underlying
measure space is resonant: this condition means that either (R, μ) is nonatomic
or it is completely atomic, all atoms having equal measure. From now on we
shall suppose that (R, μ) is nonatomic.

Given any r.i. space X over a measure space (R, μ), the Luxemburg representation theorem (see Chapter 2, Theorem 4.10 of [21]) implies that there is
a unique r.i. space $X(0, \mu(R))$ over the interval $(0, \mu(R))$ endowed with the
one-dimensional Lebesgue measure such that $\|f\|_X = \|f^*\|_{X(0,\mu(R))}$. This space
is called the *representation space* of X and is often denoted by $\overline{X}(0, \mu(R))$ or

even $\overline{X}$. When $R = (0, \infty)$ and μ is Lebesgue measure, every r.i. space X over (R, μ) coincides with its representation space.

If $X = X(\rho)$ is the r.i. space determined by an r.i. norm ρ, its *fundamental function* ϕ_X is defined by

$$\phi_X(t) = \rho(\chi_E), t \in [0, \mu(R)),$$

where $E \subset R$ is such that $\mu(E) = t$. That ϕ_X is well defined is a consequence of the properties of r.i. norms and the nonatomicity of (R, μ). A useful property of the fundamental function is that

$$\phi_X(t)\phi_{X'}(t) = t, \ t \in [0, \mu(R)).$$

When $p \in [1, \infty)$ and $X = L_p(R, \mu)$ it is easy to see that $\phi_X(t) = t^{1/p}$ $(t \in [0, \mu(R))$.

Basic examples of r.i. spaces are provided by the usual L_p spaces. With an eye to later applications, we confine ourselves for the moment to the case in which R is an open subset Ω of $\mathbb{R}^n$ and μ is Lebesgue n-measure. For $p \in [1, \infty]$ define the functional ρ_p by

$$\rho_p(f) = \|f\|_{p,\Omega} = \|f\|_p = \begin{cases} \left(\int_\Omega |f|^p \, dx\right)^{1/p}, & \text{if } 1 \le p < \infty, \\ \text{ess sup}_\Omega |f|, & \text{if } p = \infty \end{cases}$$

for $f \in \mathcal{M}(\Omega, \mu)$. This is an r.i. norm with corresponding r.i. space $L_p(\Omega)$. More generally, given $p, q \in [1, \infty]$, define $\rho_{p,q}$ by

$$\rho_{p,q}(f) = \|f\|_{p,q,\Omega} = \|f\|_{p,q} = \left\| s^{1/p-1/q} f^*(s) \right\|_{q,(0,|\Omega|)}$$

for $f \in \mathcal{M}(\Omega, \mu)$. The set $L_{p,q}(\Omega)$, defined to be the family of all $f \in \mathcal{M}(\Omega, \mu)$ such that $\rho_{p,q}(f) < \infty$, is called a *Lorentz space*. If either $1 < p < \infty$ and $1 \le q \le \infty$, or $p = q = 1$, or $p = q = \infty$, then $\rho_{p,q}$ is equivalent to an r.i. norm in the sense that there are an r.i. norm σ and a constant $C \in (0, \infty)$, depending on p and q but independent of f, such that for all $f \in L_{p,q}(\Omega)$,

$$C^{-1}\sigma(f) \le \rho_{p,q}(f) \le C\sigma(f).$$

Accordingly $L_{p,q}(\Omega)$ is considered to be an r.i. space for these values of p and q: see [21], Chapter 4. If $p = 1$ and $q > 1$, then $L_{p,q}(\Omega)$ is a quasi-normed space; if $p = \infty$ and $q < \infty$, then $L_{p,q}(\Omega) = \{0\}$.

For all $p \in [1, \infty]$, $L_{p,p}(\Omega) = L_p(\Omega)$. The dependence of the Lorentz spaces on the first index is given by

$$L_{r,s}(\Omega) \hookrightarrow L_{p,q}(\Omega) \text{ if } 1 \le p < r \le \infty \text{ and } q, s \in [1, \infty];$$

as for the second index we have, when $|\Omega| < \infty$,

$$L_{p,q}(\Omega) \hookrightarrow L_{p,r}(\Omega) \text{ if } p \in [1, \infty] \text{ and } 1 \le q < r \le \infty.$$

For further details and proofs see [60] and [146].

Given any sequence $\{E_j\}$ of measurable subsets of R, we write $E_j \to \emptyset$ a.e. if $\chi_{E_j} \to 0 \ \mu$−a.e. If every f in the Banach function space X has the property that $\left\|f\chi_{E_j}\right\|_X \to 0$ whenever $E_j \to \emptyset$ a.e., X is said to have *absolutely continuous norm*; it turns out that to verify this property it is enough to consider decreasing sequences $\{E_j\}$. When X and Y are Banach function spaces (with the same underlying measure space), we say that X is *almost compactly embedded in* Y, and write $X \overset{*}{\hookrightarrow} Y$, if, for every sequence $\{E_j\}$ of measurable sets such that $E_j \to \emptyset$ a.e., we have

$$\lim_{j\to\infty} \sup_{\|u\|_X \le 1} \left\|u\chi_{E_j}\right\|_Y = 0.$$

Replacement of $\{E_j\}$ by $\{\cup_{k\ge j} E_k\}$ shows that in this definition the sequence $\{E_j\}$ may be taken to be non-increasing. This notion is useful in establishing the compactness of embeddings of Sobolev spaces. For a connected account of it, based on results given in [159] and [76], we refer to [146]. Among the important properties established in these references are the following, in which X, Y are assumed to be Banach function spaces over a σ-finite measure space (R, μ):

 (i) If $X \hookrightarrow\hookrightarrow Y$, then $X \overset{*}{\hookrightarrow} Y$.
 (ii) $X \overset{*}{\hookrightarrow} Y$ if and only if $Y' \overset{*}{\hookrightarrow} X'$.
 (iii) $X \overset{*}{\hookrightarrow} Y$ if and only if for every sequence $\{f_k\}$ of μ−measurable functions on R satisfying $\|f_k\|_X \le 1$ and $f_k \to 0 \ \mu$−a.e. we have $\|f_k\|_Y \to 0$.
 (iv) If (R, μ) is completely atomic, then $X \overset{*}{\hookrightarrow} Y$ if and only if $X \hookrightarrow\hookrightarrow Y$.
 (v) If (R, μ) is nonatomic and $\mu(R) = \infty$, then there is no pair X, Y such that $X \overset{*}{\hookrightarrow} Y$.
 (vi) Suppose that (R, μ) is nonatomic and $0 < \mu(R) < \infty$. Then $X \overset{*}{\hookrightarrow} Y$ implies that

$$\lim_{t\to 0+} \phi_Y(t)/\phi_X(t) = 0;$$

the converse is false: see [107], p. 286. Moreover, $X \overset{*}{\hookrightarrow} Y$ if and only if

$$\lim_{t\to 0+} \sup_{\|f\|_X \le 1} \sup_{\mu(E)\le t} \|f\chi_E\|_Y = 0;$$

or equivalently,

$$\lim_{t\to 0+} \sup_{\|f\|_X \le 1} \left\|f^*\chi_{[0,t)}\right\|_{\overline{Y}} = 0.$$

From (vi) it follows immediately that if Ω is a bounded open subset of $\mathbb{R}^n$ and $1 \le q < p < \infty$, then $L_p(\Omega) \overset{*}{\hookrightarrow} L_q(\Omega)$. In fact, a similar argument shows that

$L_p(\Omega) \overset{*}{\hookrightarrow} L_{p,r}(\Omega)$ if $1 < p < r$ and $r \in (1, \infty)$, which since $L_{p,r}(\Omega) \hookrightarrow L_q(\Omega)$ if $p > q$ gives a sharpening of the earlier result.

1.4 The Palais–Smale Condition

Let X be a uniformly convex Banach space and suppose that $G \in C^1(X, \mathbb{R})$, so that the Fréchet derivative G' of G belongs to $B(X, X^*)$. A point $x \in X$ is said to be *critical* (for G) if $G'(x) = 0$; otherwise x is called *regular*. A real number λ is a *critical value* of G if there is a critical point $x \in X$ such that $G(x) = \lambda$; otherwise λ is a r*egular value* of G (even if $\lambda \notin G(X)$).

Now let $M := \{u \in X : G(u) = 1\}$; assumed to be non-empty, M is a C^1 manifold (see, for example, [51]); and we suppose that 1 is a regular value of G. Given $u \in M$,

$$T_u M := \{x \in X : \langle x, G'(u) \rangle = 0\}$$

is the *tangent space* $T_u M$ of M at u; the norm on the dual space $(T_u M)^*$ will be denoted by $\|\cdot \mid (T_u M)^*\|$. Let $\Phi \in C^1(X, \mathbb{R})$ and represent its restriction to M by $\widetilde{\Phi}$. For each $u \in M$ the norm of the derivative of $\widetilde{\Phi}$ at u is

$$\left\|\widetilde{\Phi}'(u)\right\|_* := \left\|\Phi'(u) \mid (T_u M)^*\right\|.$$

The functional Φ is said to satisfy the *Palais–Smale condition at level c* (written $\Phi \in (PS)_{c,M}$) if every sequence $\{u_j\}$ in M such that

$$\lim_{j \to \infty} \Phi(u_j) = c \text{ and } \lim_{j \to \infty} \left\|\widetilde{\Phi}'(u_j)\right\|_* = 0$$

has a convergent subsequence.

Given any $k \in \mathbb{N}$, the unit sphere of $\mathbb{R}^k$ is denoted by S^k and we write

$$C_o\left(S^k, M\right) = \left\{h \in C\left(S^k, M\right) : h \text{ is odd}\right\}.$$

Finally, we state a theorem due to Cuesta [46].

Theorem 1.8 *Let $\Phi \in C^1(X, \mathbb{R})$ be even, suppose that $k \in \mathbb{N}$, set*

$$d = \inf_{h \in C_o(S^k, M)} \max_{z \in S^k} \Phi(h(z))$$

and assume that $d \in \mathbb{R}$. If $\Phi \in (PS)_{d,M}$, then there exists $u \in M$ such that $\Phi(u) = d$ and $\widetilde{\Phi}'(u) = 0$.

The proof uses the Ekeland variational principle [71] and is too long to reproduce here. Application of this result will be made in Chapter 4 in connection with the second eigenvalue of the fractional p-Laplacian.

1.5 Inequalities

Here we collect some inequalities that will be useful later on, following largely the presentation of Appendix B of [29]. For shortness of presentation we introduce the function $J_p \colon \mathbb{R} \to [0, \infty)$ defined for each $p \in (1, \infty)$ by $J_p(t) = |t|^{p-2} t$. Note that on $[0, \infty)$ it is convex when $p > 2$ and concave if $p \in (1, 2]$.

Lemma 1.9   *Suppose that $a, b \in \mathbb{R}$ and $ab \leq 0$. Then*

$$|a - b|^{p-2} (a - b)a \geq \begin{cases} |a|^p - (p - 1)\, |a - b|^{p-2}\, ab & \text{if } p \in (1, 2], \\ |a|^p - (p - 1)\, |a|^{p-2}\, ab & \text{if } p \in (2, \infty). \end{cases}$$

Proof   We may suppose that $a \geq 0$ and $b \leq 0$. When $p > 2$, the convexity of J_p implies that

$$J_p(x) + (y - x)J_p'(x) \leq J_p(y), \quad 0 \leq x \leq y; \tag{1.5.1}$$

similarly, when $p \in (1, 2]$ we have

$$J_p(x) + (y - x)J_p'(y) \leq J_p(y), \quad 0 \leq x \leq y. \tag{1.5.2}$$

Suppose that $p \in (1, 2]$. Then from (1.5.2) with $y = a - b$ and $x = a$, we see that

$$\begin{aligned} |a - b|^{p-2} (a - b) = J_p(a - b) &\geq J_p(a) - b J_p'(a - b) \\ &= |a|^p - (p - 1)\, |a - b|^{p-2}\, b, \end{aligned}$$

from which the desired result follows immediately. The argument when $p > 2$ is similar, this time using (1.5.1).   □

Lemma 1.10   *Let $p \in (1, \infty)$. Then there is a constant $c = c(p) > 0$ such that for all $a, b \in \mathbb{R}$,*

$$|a - b|^p \leq |a|^p + |b|^p + c \left(|a|^2 + |b|^2 \right)^{(p-2)/2} |ab|.$$

Proof   When $ab \geq 0$ it is enough to suppose that $a, b \geq 0$ and $a \geq b$. But then

$$|a - b|^p \leq a^p \leq |a|^p + |b|^p$$

and the result follows. On the other hand, if $ab \leq 0$, then we may assume that $a \geq 0$ and $b \leq 0$, so that $b = -d$ for some $d \geq 0$. We have to show that

$$(a + d)^p \leq a^p + d^p + c \left(a^2 + d^2 \right)^{(p-2)/2} ad.$$

This is obvious when $a = 0$, and so it is enough to prove that

$$(1 + x)^p \leq 1 + x^p + c \left(1 + x^2 \right)^{(p-2)/2} x.$$

Since

$$\lim_{x \to 0+} \frac{(1+x)^p - 1 - x^p}{\left(1+x^2\right)^{(p-2)/2} x} = \lim_{x \to \infty} \frac{(1+x)^p - 1 - x^p}{\left(1+x^2\right)^{(p-2)/2} x} = p,$$

the claim follows and completes the proof. ☐

Lemma 1.11 *For all* $a, b \in \mathbb{R}$,

$$\left(|b|^{p-2} b - |a|^{p-2} a\right) (b-a) \geq \begin{cases} (p-1) |b-a|^2 \left(|a|^2 + |b|^2\right)^{-(2-p)/2} & \text{if } p \in (1, 2], \\ 2^{2-p} |b-a|^p & \text{if } p \in (2, \infty). \end{cases}$$

Proof Suppose that $p \in (1, 2]$. Then

$$\left(J_p(b) - J_p(a)\right) (b - a) = (b - a) \int_0^1 \frac{d}{dt} J_p \left((1 - t)a + tb\right) dt$$

$$= (p - 1)(b - a)^2 \int_0^1 |(1 - t)a + tb|^{p-2} dt.$$

Since

$$|(1 - t)a + tb|^2 \leq (1 - t) |a|^2 + t |b|^2 \leq |a|^2 + |b|^2,$$

the desired inequality follows.

When $p \in (2, \infty)$ we argue as in [128] and use the identity

$$\left(|b|^{p-2} b - |a|^{p-2} a\right) (b - a) = \frac{|b|^{p-2} + |a|^{p-2}}{2} |b - a|^2$$

$$+ \frac{\left(|b|^{p-2} - |a|^{p-2}\right) \left(|b|^2 - |a^2|\right)}{2},$$

from which it is immediate that

$$\left(|b|^{p-2} b - |a|^{p-2} a\right) (b - a) \geq 2^{-1} \left(|b|^{p-2} + |a|^{p-2}\right) |b - a|^2$$

$$\geq 2^{2-p} |b - a|^p.$$

☐

2

Classical Sobolev Spaces

2.1 Basic Definitions

Let $p \in [1, \infty]$ and $k, n \in \mathbb{N}$; suppose that Ω is a non-empty open subset of $\mathbb{R}^n$. Then

$$W_p^k(\Omega) := \left\{ u \colon D^\alpha u \in L_p(\Omega) \text{ for all } \alpha \in \mathbb{N}_0^n \text{ with } |\alpha| \leq k \right\} \qquad (2.1.1)$$

is the classical Sobolev space of order k, based on $L_p(\Omega)$; the derivatives $D^\alpha u$ are taken in the sense of distributions. It is a linear space when endowed with addition and multiplication by scalars in the natural way; provided with the norm $\|\cdot\|_{k,p,\Omega}$ (written as $\|\cdot\|_{k,p}$ if there is no ambiguity) defined by

$$\|u\|_{k,p,\Omega} = \left(\sum_{|\alpha| \leq k} \|D^\alpha u\|_{p,\Omega}^p \right)^{1/p} \quad \text{if } p < \infty, \qquad (2.1.2)$$

and

$$\|u\|_{k,\infty,\Omega} = \sum_{|\alpha| \leq k} \|D^\alpha u\|_{\infty,\Omega} \qquad (2.1.3)$$

when $p = \infty$, it is a Banach space that is uniformly convex if $p \in (1, \infty)$. Moreover, $W_2^k(\Omega)$ is a Hilbert space, with inner product $(\cdot, \cdot)_{k,2,\Omega}$ (or $(\cdot, \cdot)_{k,2}$) given by

$$(u, v)_{k,2,\Omega} := \int_\Omega \sum_{|\alpha| \leq k} (D^\alpha u) \overline{(D^\alpha v)} \, dx. \qquad (2.1.4)$$

First suppose that $\Omega = \mathbb{R}^n$. In this case, characterisations of these spaces by means of the Fourier transform F are possible and desirable. To explain this, first define the function $w_s \colon \mathbb{R}^n \to \mathbb{R}$ by

$$w_s(x) = \left(1 + |x|^2\right)^{s/2} \text{ for all } x \in \mathbb{R}^n \text{ and } s \in \mathbb{R}, \qquad (2.1.5)$$

and introduce the space

$$H_p^s (\mathbb{R}^n) := \left\{ u \in \mathcal{S}' (\mathbb{R}^n) : F^{-1} (w_s Fu) \in L_p (\mathbb{R}^n) \right\}, \qquad (2.1.6)$$

where $\mathcal{S} = \mathcal{S} (\mathbb{R}^n)$ is the Schwartz space of rapidly decreasing functions and $\mathcal{S}' (\mathbb{R}^n)$ is its dual, the space of tempered distributions. Endowed with the norm

$$\left\| u | H_p^s (\mathbb{R}^n) \right\| := \left\| F^{-1} (w_s Fu) \right\|_{p, \mathbb{R}^n}, \qquad (2.1.7)$$

it is a Banach space. In fact,

$$H_p^s (\mathbb{R}^n) = W_p^s (\mathbb{R}^n) \text{ when } p \in (1, \infty) \text{ and } s \in \mathbb{N}, \qquad (2.1.8)$$

with equivalent norms. When $p = 2$ this assertion is an immediate consequence of the fact that the Fourier transform F and its inverse are unitary operators in $L_2 (\mathbb{R}^n)$; for other values of p appeal to the Michlin–Hörmander Fourier multiplier theorem gives the result. We refer to [95], 3.6.1 for further details of the argument.

Now let Ω be a non-empty open subset of $\mathbb{R}^n$. In addition to the 'intrinsic' definition of $W_p^k (\Omega)$ described above it would be natural to define this space as

$$\left\{ u \in L_p (\Omega) : \text{ there exists } v \in W_p^k (\mathbb{R}^n) \text{ with } v|_\Omega = u \right\}$$

and give it the norm

$$\inf \left\{ \|v\|_{k, p, \mathbb{R}^n} : v \in W_p^k (\mathbb{R}^n), \ v|_\Omega = u \right\},$$

where $v|_\Omega = u$ is meant in the sense of $\mathcal{D}'(\Omega)$, so that $v(\phi) = u(\phi)$ for all $\phi \in \mathcal{D} (\Omega)$. This space, defined by restriction, coincides with $W_p^k (\Omega)$ if Ω is bounded and has a Lipschitz boundary; without some condition on the boundary the spaces may be different. Similarly, $H_p^k (\Omega)$ may be defined by restriction of elements of $H_p^k (\mathbb{R}^n)$, and coincides with $W_p^k (\Omega)$ when $p \in (1, \infty)$ and $k \in \mathbb{N}$, if Ω is bounded and has sufficiently smooth boundary. It is common to denote $H_2^k (\Omega)$ by $H^k (\Omega)$. More information on this topic is given in [95], Chapters 3 and 4.

2.2 Fundamental Results

Here we list, for ease of reference, some of the most useful results concerning Sobolev spaces. In what follows the closure of $C_0^\infty (\Omega)$ in $W_p^k (\Omega)$ will be denoted by $\overset{0}{W}{}_p^k (\Omega)$. Proofs may be found in [64] and [61]; see also [32] and [141]. We begin with embeddings.

Theorem 2.1 *Let Ω be a bounded open subset of $\mathbb{R}^n$, let $k \in \mathbb{N}$ and suppose that $p \in [1, \infty)$.*

(i) *Assume that Ω has Lipschitz boundary and that $kp < n$. Then*

$$W_p^k(\Omega) \hookrightarrow L_s(\Omega) \; if \; s \in \left[p, np/(n-kp)\right],$$

and this embedding is compact if $s \in [p, np/(n-kp))$. If for some $l \in \mathbb{N}_0$ and $\gamma \in (0,1]$ the inequality $(k-l-\gamma)p \geq n$ holds, then

$$W_p^k(\Omega) \hookrightarrow C^{l,\gamma}\left(\overline{\Omega}\right),$$

and the embedding is compact if $(k-l-\gamma)p > n$. These results hold without any condition on $\partial\Omega$ if $W_p^k(\Omega)$ is replaced by $\overset{0}{W_p^k}(\Omega)$.

(ii) *If $k, l \in \mathbb{N}_0$, $l > k$ and $\partial\Omega$ is of class C, then $W_p^l(\Omega) \hookrightarrow\hookrightarrow W_p^k(\Omega)$; the condition on $\partial\Omega$ may be dropped if $W_p^l(\Omega)$ and $W_p^k(\Omega)$ are replaced by $\overset{0}{W_p^l}(\Omega)$ and $\overset{0}{W_p^k}(\Omega)$, respectively.*

(iii) *If $p \in (1,\infty)$, then $W_p^k(\Omega) \hookrightarrow\hookrightarrow W_q^{k-1}(\Omega)$ whenever $q \in [1,p)$. Note that no condition on $\partial\Omega$ is required.*

The theorem shows that the embedding $I_{p,q}$ of $W_p^1(\Omega)$ in $L_q(\Omega)$ is compact whenever $p \in (1,\infty)$ and $q \in [1,p)$, no matter how unpleasant $\partial\Omega$ may be. There is a dramatic change when $q = p$, for while $I_{p,p}$ is compact when $\partial\Omega$ is of class C, the 'rooms and passages' example (see Theorem V.4.18 of [64]) shows that in the absence of any condition on the boundary of Ω, $I_{p,p}$ may be noncompact. Indeed, [68] contains an example in which $I_{p,p}$ is not even strictly singular.

It turns out that these results may be refined by use of the Lorentz spaces introduced in Section 1.3.3. For example, as regards (i), it can be shown that when $1 \leq p < n$, the smallest r.i. space $X(\Omega)$ such that $\overset{0}{W_p^1}(\Omega) \hookrightarrow X(\Omega)$ is $L_{p^*,p}(\Omega)$, where $p^* = np/(n-p)$, so that the embedding $\overset{0}{W_p^1}(\Omega) \hookrightarrow L_{p^*,p}(\Omega)$ is optimal in the class of r.i. spaces, so far as the target space is concerned. The more complicated question of optimality of the domain space in such embeddings is briefly discussed in p. 23 of [65], where references to further work on this topic can be found.

Next we list some very useful inequalities. We say that an open set $\Omega \subset \mathbb{R}^n$ supports the *p-Friedrichs inequality* if there is a constant $c > 0$ such that for all $u \in C_0^\infty(\Omega)$,

$$\|u\|_{p,\Omega} \leq c \, \||\nabla u|\|_{p,\Omega}.$$

An example of such a set is given in the next result: note that the set Ω need not be bounded.

Theorem 2.2 *Let Ω be an open subset of $\mathbb{R}^n$ that lies between two parallel co-ordinate hyperplanes at a distance l apart, and suppose that $p \in [1, \infty)$. Then for all $u \in \overset{0}{W}{}^1_p (\Omega)$,*

$$\|u\|_{p,\Omega} \leq l \, \||\nabla u|\|_{p,\Omega} \, .$$

If the parallel hyperplanes are not parallel to co-ordinate hyperplanes the inequality still holds but with l replaced by Cl for some constant C independent of u. When Ω is bounded we have the following.

Theorem 2.3 *Let Ω be a bounded open subset of $\mathbb{R}^n$ and suppose that $p \in [1, \infty]$. Then for all $u \in \overset{0}{W}{}^1_p (\Omega)$,*

$$\|u\|_{p,\Omega} \leq (|\Omega| / \omega_n)^{1/n} \, \||\nabla u|\|_{p,\Omega} \, .$$

If $p \in [1, n)$ and $q \in \left[p, np/(n-p)\right]$, then there is a constant C such that for all $u \in \overset{0}{W}{}^1_p (\Omega)$,

$$\|u\|_{q,\Omega} \leq C \, |\Omega|^{1/n + 1/q - 1/p} \, \||\nabla u|\|_{p,\Omega} \, .$$

It is sometimes convenient to use the *homogeneous Sobolev space* $\overset{0}{\mathcal{D}}{}^1_p (\Omega)$, where

$$\overset{0}{\mathcal{D}}{}^1_p (\Omega) := \text{completion of } C^\infty_0 (\Omega) \text{ with respect to the norm } u \longmapsto \||\nabla u|\|_{p,\Omega} \, .$$

This coincides with $\overset{0}{W}{}^1_p (\Omega)$ when Ω supports the *p-Friedrichs inequality*, in which case

$$\overset{0}{\mathcal{D}}{}^1_p (\Omega) \subset \left\{ u \in W^1_p (\mathbb{R}^n) : u = 0 \text{ a.e. in } \mathbb{R}^n \backslash \Omega \right\},$$

with equality when $\partial \Omega$ is of class C (see [90], Theorem 1.4.2.2). For elements of the whole space $W^1_p (\Omega)$ there is

Theorem 2.4 (The Poincaré inequality) *Let Ω be a bounded, convex open subset of $\mathbb{R}^n$ with diameter d and let $p \in [1, \infty]$. Then for all $u \in W^1_p (\Omega)$,*

$$\|u - u_\Omega\|_{p,\Omega} \leq (\omega_n / |\Omega|)^{1-1/n} d^n \, \||\nabla u|\|_{p,\Omega},$$

where $u_\Omega = |\Omega|^{-1} \int_\Omega u(x) \, dx$.

The following results concerning $\overset{0}{W}{}^1_p (\Omega)$ are useful.

Theorem 2.5 *Let Ω be an open subset of $\mathbb{R}^n$ and let $p \in (1, \infty)$.*

(i) If $u \in W^1_p (\Omega)$ and supp u is a compact subset of Ω, then $u \in \overset{0}{W}{}^1_p (\Omega)$.

(ii) Suppose that $u \in W_p^1(\Omega) \cap C(\overline{\Omega})$. If $u = 0$ on $\partial\Omega$, $u \in \overset{0}{W_p^1}(\Omega)$; if $\partial\Omega \in C^1$ and $u \in \overset{0}{W_p^1}(\Omega)$, then $u = 0$ on $\partial\Omega$.

Next we give some details of the behaviour under translation of functions in $W_p^1(\Omega)$ that will be used extensively later on.

Proposition 2.6 *Let Ω be an open subset of $\mathbb{R}^n$, let $p \in (1, \infty)$ and suppose that $u \in W_p^1(\Omega)$. Then for every open subset ω of $\mathbb{R}^n$ with compact closure contained in Ω, and all $h \in \mathbb{R}^n$ with $|h| < \text{dist}(\omega, \partial\Omega)$,*

$$\int_\omega |u(x+h) - u(x)|^p \, dx \le |h|^p \, \||\nabla u|\|_{p,\Omega}^p . \tag{2.2.1}$$

If $\Omega = \mathbb{R}^n$, then for all $h \in \mathbb{R}^n$,

$$\int_{\mathbb{R}^n} |u(x+h) - u(x)|^p \, dx \le |h|^p \, \||\nabla u|\|_{p,\mathbb{R}^n}^p . \tag{2.2.2}$$

Proof   First suppose that $u \in C_0^\infty(\mathbb{R}^n)$. Let $h, x \in \mathbb{R}^n$ and set $v(t) = u(x + th)$ $(t \in \mathbb{R})$. Then $v'(t) = h \cdot \nabla u(x + th)$ and

$$u(x+h) - u(x) = v(1) - v(0) = \int_0^1 h \cdot \nabla u(x+th) \, dt,$$

so that

$$|u(x+h) - u(x)|^p \le |h|^p \int_0^1 |\nabla u(x+th)|^p \, dt. \tag{2.2.3}$$

Thus

$$\int_\omega |u(x+h) - u(x)|^p \, dx \le |h|^p \int_\omega dx \int_0^1 |\nabla u(x+th)|^p \, dt$$
$$= |h|^p \int_0^1 dt \int_{\omega+th} |\nabla u(y)|^p \, dy.$$

If $|h| < \text{dist}(\omega, \partial\Omega)$, there is an open subset ω' of $\mathbb{R}^n$ with compact closure contained in Ω such that $\omega + th \subset \omega'$ for all $t \in [0, 1]$. Hence

$$\int_\omega |u(x+h) - u(x)|^p \, dx \le |h|^p \int_{\omega'} |\nabla u(y)|^p \, dy, \tag{2.2.4}$$

which gives (2.2.1) when $u \in C_0^\infty(\mathbb{R}^n)$. If $u \in W_p^1(\Omega)$, then by Theorem 1.3.14 of [65], there is a sequence $\{u_k\}$ of functions in $C_0^\infty(\mathbb{R}^n)$ such that as $k \to \infty$, $u_k \to u$ in $L_p(\Omega)$ and $\nabla u_k \to \nabla u$ in $(L_p(\omega'))^n$ for all ω' with compact closure contained in Ω. Application of (2.2.4) to (u_k) and letting $k \to \infty$ gives (2.2.1) for all $u \in W_p^1(\Omega)$; and (2.2.2) follows immediately. $\square$

To conclude this chapter, we give the result that the composition of a bounded Sobolev embedding with an almost compact embedding is compact, following the presentation of Slavíková [159]. Given an open subset Ω of $\mathbb{R}^n$, suppose that X is a Banach function space on (Ω, μ_n), where μ_n is a Lebesgue n-measure, and let

$$W^1 X := \{f \in X \colon |\nabla f| \in X\},$$

where the functions involved are real-valued and weakly differentiable. We endow $W^1 X$ with the norm $\|\cdot\|_X + \||\nabla \cdot|\|_X$. For example, when $X = L_p(\Omega)$ the space just constructed is simply the usual Sobolev space $W_p^1(\Omega)$.

Theorem 2.7 *Let Ω be an open subset of R^n and suppose that X, Y, Z are Banach function spaces over (Ω, μ_n) such that $W^1 X \hookrightarrow Y$ and $Y \overset{*}{\hookrightarrow} Z$. Then $W^1 X \hookrightarrow\hookrightarrow Z$.*

In Chapter 3 we give an adaptation of this result of Slavíková to deal with fractional Sobolev spaces.

Further illustrations and consequences of this line of thought are given in [145]. In [67], Theorem 2.7 is used to obtain concrete conditions sufficient to ensure the compactness of embeddings of Sobolev spaces based on spaces with variable exponent.

3

Fractional Sobolev Spaces

3.1 Definitions

Let Ω be a non-empty open subset of $\mathbb{R}^n$, suppose that $p \in [1, \infty)$ and let $s > 0$, $s \notin \mathbb{N}$. The so-called fractional Sobolev space $W_p^s(\Omega)$ arose in an attempt to fill the gaps between $L_p(\Omega)$, $W_p^1(\Omega)$, $W_p^2(\Omega)$, ...; it was introduced independently and more or less simultaneously by Aronszajn [13], Gagliardo [86] and Slobodeckij [161]. We begin by dealing with the case in which $s \in (0, 1)$. Then

$$W_p^s(\Omega) := \left\{ u \in L_p(\Omega) \colon (x, y) \longmapsto \frac{|u(x) - u(y)|}{|x - y|^{\frac{n}{p} + s}} \in L_p(\Omega \times \Omega) \right\};$$

endowed with the norm

$$\left\| u \mid W_p^s(\Omega) \right\|_{s,p,\Omega} := \left(\int_\Omega |u(x)|^p \, dx + \int_\Omega \int_\Omega \frac{|u(x) - u(y)|^p}{|x - y|^{n+sp}} \, dx \, dy \right)^{1/p},$$

it is a Banach space. Associated with this norm is the *Gagliardo seminorm*

$$[u]_{s,p,\Omega} := \left(\int_\Omega \int_\Omega \frac{|u(x) - u(y)|^p}{|x - y|^{n+sp}} \, dx \, dy \right)^{1/p}.$$

If $1 < p < \infty$, the space $W_p^s(\Omega)$ is reflexive. To establish this, define

$$T \colon W_p^s(\Omega) \to L_p(\Omega) \times L_p(\Omega \times \Omega) := E$$

by $Tu = (u, U)$, where

$$U(x, y) = \frac{u(x) - u(y)}{|x - y|^{\frac{n}{p} + s}}.$$

When E is furnished with the norm

$$\|(u, v)\|_E := \left(\|u\|_{p,\Omega}^p + \|v\|_{p,\Omega \times \Omega}^p \right)^{1/p},$$

it is reflexive; since T is a linear isometry, $T\left(W_p^s(\Omega)\right)$ is a closed subspace of E and is therefore also reflexive. The reflexivity of $W_p^s(\Omega)$ follows; in fact, the same argument establishes uniform convexity and uniform smoothness of this space.

Higher-order fractional spaces are introduced in a natural way: suppose that $s = k + \sigma$, where $k \in \mathbb{N}_0$ and $\sigma \in (0, 1)$. Then

$$W_p^s(\Omega) := \left\{u \in W_p^k(\Omega) : D^\alpha u \in W_p^\sigma(\Omega) \text{ for all } \alpha \in \mathbb{N}_0^n \text{ with } |\alpha| = k\right\};$$

equipped with the norm

$$\left\|u | W_p^s(\Omega)\right\| := \left(\left\|u | W_p^k(\Omega)\right\|^p + \sum_{|\alpha|=k} \left\|D^\alpha u | W_p^\sigma(\Omega)\right\|^p\right)^{1/p},$$

it becomes a Banach space; it is reflexive if $1 < p < \infty$.

Another attempt to fill the gaps between the classical spaces is provided, when $\Omega = \mathbb{R}^n$, by the spaces $H_p^s(\mathbb{R}^n)$ defined via the Fourier transform by (2.1.6) above. As we have seen,

$$H_p^k(\mathbb{R}^n) = W_p^k(\mathbb{R}^n) \text{ if } 1 < p < \infty \text{ and } k \in \mathbb{N}_0,$$

but (see [95], p. 82) if $s > 0$, $s \notin \mathbb{N}$ and $1 < p < \infty$, then

$$H_p^s(\mathbb{R}^n) = W_p^s(\mathbb{R}^n) \text{ if and only if } p = 2.$$

Moreover, $H_p^s(\mathbb{R}^n)$ and $W_p^s(\mathbb{R}^n)$ are isomorphic if and only if $p = 2$: see [95], pp. 84–85. These facts suggest that the most natural extension of the classical Sobolev spaces involving an arbitrary smoothness parameter s is not $W_p^s(\mathbb{R}^n)$ but $H_p^s(\mathbb{R}^n)$. However, the more explicit definition of the norm on $W_p^s(\mathbb{R}^n)$ has advantages, notably in connection with the description of trace spaces associated with the restrictions to hyperplanes in $\mathbb{R}^n$ of functions belonging to Sobolev spaces, and also in the determination of optimal constants in various inequalities.

3.2 Basic Properties

We begin with an illustration of the effect of change of the smoothness parameter s.

Proposition 3.1 *Let Ω be an open subset of $\mathbb{R}^n$ and suppose that $p \in [1, \infty)$.*

(i) If $0 < s_1 \leq s_2 < 1$, then $W_p^{s_2}(\Omega) \hookrightarrow W_p^{s_1}(\Omega)$.

(ii) *If Ω has a bounded Lipschitz boundary or $\Omega = \mathbb{R}^n$, then $W_p^1(\Omega) \hookrightarrow W_p^s(\Omega)$ for all $s \in (0, 1)$.*

Proof   To deal with (i), let $u \in W_p^{s_2}(\Omega)$. Then

$$\int_\Omega \int_{\Omega \cap \{|x-y| \geq 1\}} \frac{|u(x)|^p}{|x-y|^{n+s_1 p}} \, dx \, dy \leq \int_\Omega \left(\int_{|z| \geq 1} \frac{dz}{|z|^{n+s_1 p}} \right) |u(x)|^p \, dx$$

$$= \|u\|_{p,\Omega}^p \, \omega_{n-1}/(s_1 p).$$

Hence

$$\int_\Omega \int_{\Omega \cap \{|x-y| \geq 1\}} \frac{|u(x) - u(y)|^p}{|x - y|^{n+s_1 p}} \, dx \, dy$$

$$\leq 2^{p-1} \int_\Omega \int_{\Omega \cap \{|x-y| \geq 1\}} \frac{|u(x)|^p + |u(y)|^p}{|x - y|^{n+s_1 p}} \, dx \, dy$$

$$\leq 2^p \|u\|_{p,\Omega}^p \, \omega_{n-1}/(s_1 p). \tag{3.2.1}$$

Moreover,

$$\int_\Omega \int_{\Omega \cap \{|x-y| < 1\}} \frac{|u(x) - u(y)|^p}{|x - y|^{n+s_1 p}} \, dx \, dy \leq \int_\Omega \int_{\Omega \cap \{|x-y| < 1\}} \frac{|u(x) - u(y)|^p}{|x - y|^{n+s_2 p}} \, dx \, dy,$$

and so

$$\int_\Omega \int_\Omega \frac{|u(x) - u(y)|^p}{|x-y|^{n+s_1 p}} \, dx \, dy \leq 2^p \|u\|_{p,\Omega}^p \, \omega_{n-1}/(s_1 p) + \int_\Omega \int_\Omega \frac{|u(x) - u(y)|^p}{|x - y|^{n+s_2 p}} \, dx \, dy,$$

from which (i) follows.

As for (ii), given $u \in W_p^1(\Omega)$, in view of the assumptions on $\partial \Omega$ there is an extension $\tilde{u} \in W_p^1(\mathbb{R}^n)$ of u such that $\|\tilde{u}\|_{1,p,\mathbb{R}^n} \leq C \|u\|_{1,p,\Omega}$ for some constant C independent of u (see [64], V.4). This extension is, of course, not needed if $\Omega = \mathbb{R}^n$). With $z = y - x$ and B representing the unit ball centred at 0 we have (see (2.2.3))

$$\int_{\Omega} \int_{\Omega \cap \{|x-y|<1\}} \frac{|u(x) - u(y)|^p}{|x-y|^{n+sp}} \, dx \, dy$$

$$\leq \int_{\Omega} \int_{B} \frac{|u(x) - u(z+x)|^p}{|z|^{n+sp}} \, dz \, dx$$

$$= \int_{\Omega} \int_{B} \frac{|u(x) - u(z+x)|^p}{|z|^p} \cdot \frac{1}{|z|^{n+(s-1)p}} \, dz \, dx$$

$$\leq \int_{\Omega} \int_{B} \left(\int_{0}^{1} \frac{|\nabla u(x+tz)|}{|z|^{\frac{n}{p}+s-1}} dt \right)^p dz \, dx$$

$$\leq \int_{\mathbb{R}^n} \int_{B} \int_{0}^{1} \frac{|\nabla \tilde{u}(x+tz)|^p}{|z|^{n+p(s-1)}} \, dt \, dz \, dx$$

$$\leq \int_{B} \int_{0}^{1} \frac{\|\nabla \tilde{u}\|_{p,\mathbb{R}^n}^p}{|z|^{n+p(s-1)}} \, dt \, dz,$$

which is bounded above by a constant multiple of $\|u\|_{1,p,\Omega}^p$. Together with (3.2.1) this gives the result. $\qquad \square$

Remark 3.2 The necessity of some condition on $\partial\Omega$ for (ii) to hold is illustrated by an example in Section 9 of [142], where it is shown that given any $s \in (0,1)$, there exist $p \in (1/s, \infty)$ and an open set $\Omega \subset \mathbb{R}^2$, with boundary of cusp type, such that $W_p^1(\Omega) \subsetneq W_p^s(\Omega)$. More precisely, let

$$\Omega = (\mathbb{R}^2 \setminus \mathcal{C}) \cap B(0,1), \quad \text{where } \mathcal{C} = \{(x_1, x_2) : x_1 \leq 0, |x_2| \leq |x_1|^\kappa\},$$

where $\kappa > (p+1)/(p-1)$; describe points $x = (x_1, x_2) \in \mathbb{R}^2 \setminus \mathcal{C}$ by polar co-ordinates $\rho(x) \in (0, \infty)$ and $\theta(x) \in (-\pi, \pi)$, and define $u(x) = \rho(x)\theta(x)$. Computations then show that $u \in W_p^1(\Omega) \setminus W_p^s(\Omega)$.

Corollary 3.3 *Let Ω be an open subset of $\mathbb{R}^n$ with bounded Lipschitz boundary, suppose that $p \in [1, \infty)$ and let $s_1, s_2 > 1$, with $s_2 \geq s_1$. Then $W_p^{s_2}(\Omega) \hookrightarrow W_p^{s_1}(\Omega)$.*

Proof The result is clear if $s_1, s_2 \in \mathbb{N}$. Suppose that $s_i = k_i + \sigma_i$, with $k_i \in \mathbb{N}$ and $\sigma_i \in (0,1)$ $(i = 1, 2)$. If $k_1 = k_2$, then the claim follows from Proposition 3.1 (i). If $k_2 \geq k_1 + 1$, then use of (i) and (ii) of Proposition 3.1 gives

$$W_p^{s_2}(\Omega) \hookrightarrow W_p^{k_2}(\Omega) \hookrightarrow W_p^{k_1+1}(\Omega) \hookrightarrow W_p^{k_1+\sigma_1}(\Omega).$$

The remaining case, in which exactly one of k_1, k_2 is an integer, is handled in a similar fashion. $\qquad \square$

Another example of an embedding of a Sobolev space with no condition imposed on the boundary can be obtained by using the following fractional version of Theorem 2.7.

Theorem 3.4 *Let Ω be an open subset of $\mathbb{R}^n$, suppose that $s \in (0, 1)$ and let $p \in (1, \infty)$; assume that Y, Z are Banach function spaces over Ω (endowed with Lebesgue n-measure) such that $W_p^s (\Omega) \hookrightarrow Y$ and $Y \overset{*}{\hookrightarrow} Z$. Then $W_p^s (\Omega) \hookrightarrow\hookrightarrow Z$.*

Proof   This is essentially the same as that given by Slavíková in the non-fractional case, but for the reader's convenience we give some details. Let $\{B_k\}$ be a sequence of balls contained in Ω that covers Ω and let $\{g_k\}$ be a bounded sequence in $W_p^s (\Omega)$. We claim that for each $m \in \mathbb{N}$, there is a subsequence $\left\{g_k^m\right\}_{k=1}^{\infty}$ of $\left\{g_k^{m-1}\right\}_{k=1}^{\infty}$ (with $g_k^0 = g_k$) that converges a.e. on B_m. To establish this, suppose that the sequence $\left\{g_k^{m-1}\right\}_{k=1}^{\infty}$ is known for some fixed m. Then $\left\{g_k^{m-1}\right\}_{k=1}^{\infty}$ is bounded in $W_p^s (B_m)$, and so, since $W_p^s (B_m)$ is compactly embedded in $L_p (B_m)$ (because B_m is bounded and has smooth boundary and so the fractional space can be identified with a Besov space for which this compact embedding is known), there is a subsequence of $\left\{g_k^{m-1}\right\}_{k=1}^{\infty}$ that converges in $L_p (B_m)$. There is therefore another subsequence, which we denote by $\left\{g_k^m\right\}_{k=1}^{\infty}$, that converges a.e. on B_m, and the inductive step is complete. The diagonal sequence $\left\{g_m^m\right\}_{m=1}^{\infty}$ converges a.e. on Ω to some function, g say: since $W_p^s (\Omega) \hookrightarrow Y$ it follows that $\left\{g_m^m\right\}_{m=1}^{\infty}$ is bounded in Y: by the Fatou lemma for Banach function spaces (see, for example, [146], Lemma 6.1.12),

$$\|g\|_Y \leq \liminf_{m \to \infty} \left\| g_m^m \right\|_Y < \infty.$$

Hence $g \in Y$; by property (iii) of Section 1.3.3 (characterising almost compactness) we have $\left\| g_m^m - g \right\|_Z \to 0$. Hence $W_p^s (\Omega) \hookrightarrow\hookrightarrow Z$. $\square$

The example we have in mind arises when we take Ω to be bounded, $Y = L_p(\Omega)$ and $Z = L_q(\Omega)$, where $q \in [1, p)$. Since $L_p(\Omega) \overset{*}{\hookrightarrow} L_q(\Omega)$ (see Section 1.3.3), the theorem implies that $W_p^s (\Omega) \hookrightarrow\hookrightarrow L_q(\Omega)$. This is a fractional extension of [64], Theorem V. 4.16, for bounded Ω and $k = 1$. Note that no condition on $\partial\Omega$ is required.

Let $\overset{0}{W}_p^s(\Omega)$ denote the closure of $C_0^\infty (\Omega)$ in $W_p^s(\Omega)$. In general this is a proper closed subspace of $W_p^s (\Omega)$, but

$$\overset{0}{W}_p^s (\mathbb{R}^n) = W_p^s (\mathbb{R}^n) \quad \text{for all } s > 0. \tag{3.2.2}$$

For a proof we refer to [2], Theorem 7.38.

It is often convenient to work, not with $\overset{0}{W}_p^s (\Omega)$ as defined above, but with the space $\overset{0}{X}_p^s(\Omega)$, where

$$\overset{0}{X}_p^s(\Omega) := \text{completion of } C_0^\infty(\Omega) \text{ with respect to the norm } [\cdot]_{s,p,\mathbb{R}^n} + \|\cdot\|_{p,\Omega} \,.$$

Plainly $\overset{0}{X}{}_p^s(\Omega) \subset \overset{0}{W}{}_p^s(\Omega)$. Note that if $u \in C_0^\infty(\Omega)$, then

$$\int_{\mathbb{R}^n} \int_{\mathbb{R}^n} \frac{|u(x) - u(y)|^p}{|x - y|^{n+sp}} \, dx \, dy = \int_\Omega \int_\Omega \frac{|u(x) - u(y)|^p}{|x - y|^{n+sp}} \, dx \, dy$$
$$+ 2 \int_\Omega \int_{\mathbb{R}^n \setminus \Omega} \frac{|u(x)|^p}{|x - y|^{n+sp}} \, dx \, dy,$$

that is,

$$[u]_{s,p,\mathbb{R}^n}^p = [u]_{s,p,\Omega}^p + 2 \int_\Omega \int_{\mathbb{R}^n \setminus \Omega} \frac{|u(x)|^p}{|x - y|^{n+sp}} \, dx \, dy;$$

the last term need not be zero, and might even be infinite even though supp $u \subset \Omega$; below we show that this cannot happen if the boundary of Ω is smooth enough. First we give an inequality of Friedrichs type: we say that, given $s \in (0, 1)$ and $p \in [1, \infty)$, the open set $\Omega \subset \mathbb{R}^n$ *supports the* (s, p)-*Friedrichs inequality* if there is a positive constant c such that for all $u \in C_0^\infty(\Omega)$,

$$\|u\|_{p,\Omega}^p \le c \, [u]_{s,p,\mathbb{R}^n}^p .$$

The next Proposition (given in [28]) shows that every bounded open set Ω has this property.

Proposition 3.5 *Let $p \in [1, \infty)$, $s \in (0, 1)$ and suppose that Ω is a bounded open subset of $\mathbb{R}^n$. Then for every $u \in C_0^\infty(\Omega)$,*

$$\|u\|_{p,\Omega}^p \le C \, [u]_{s,p,\mathbb{R}^n}^p ,$$

where

$$C = C(n, s, p, \Omega) = \min \left\{ \frac{\text{diam} \, (\Omega \cup B)^{n+sp}}{|B|} : B \subset \mathbb{R}^n \setminus \Omega \text{ is a ball} \right\} .$$

Proof Let $u \in C_0^\infty(\Omega)$ and let B_R be a ball of radius R contained in $\mathbb{R}^n \setminus \Omega$. For all $x \in \Omega$ and $y \in B_R$,

$$|u(x)|^p = \frac{|u(x) - u(y)|^p}{|x - y|^{n+sp}} |x - y|^{n+sp} ,$$

which gives

$$|B_R| \, |u(x)|^p \le \left(\sup_{z \in \Omega, y \in B_R} |z - y|^{n+sp} \right) \int_{B_R} \frac{|u(x) - u(y)|^p}{|x - y|^{n+sp}} \, dy.$$

The result follows. $\qquad\qquad\qquad\qquad\qquad\qquad\qquad\qquad\qquad\qquad\qquad$ $\square$

From this it is clear that when Ω is bounded the space $\overset{0}{X}{}_p^s(\Omega)$ can be equivalently defined as the completion of $C_0^\infty(\Omega)$ with respect to the norm $[\cdot]_{s,p,\mathbb{R}^n}$. The case in which Ω is contained between parallel hyperplanes, and may be unbounded, is discussed in [25], Remark 1.6; see also [42] and the references contained in that paper.

The following Proposition (see, for example, [28], Proposition B.1) implies that for $\overset{0}{X}{}^s_p(\Omega)$ to coincide with $\overset{0}{W}{}^s_p(\Omega)$ it is sufficient that $ps \neq 1$ and that Ω should be bounded and have a Lipschitz boundary.

Proposition 3.6 *Let $s \in (0, 1)$ and $p \in (1, \infty)$ be such that $ps \neq 1$; suppose Ω is a bounded open subset of $\mathbb{R}^n$ with Lipschitz boundary. Then there is a positive constant $C = C(n, p, s, \Omega)$ such that for all $u \in C^\infty_0(\Omega)$,*

$$[u]_{s,p,\mathbb{R}^n} + \|u\|_{p,\Omega} \leq C \left([u]_{s,p,\Omega} + \|u\|_{p,\Omega} \right).$$

Proof Let $u \in C^\infty_0(\Omega)$ and put $\delta(x) = \inf_{y \in \mathbb{R}^n \setminus \Omega} |x - y|$ $(x \in \Omega)$. Then

$$\mathbb{R}^n \setminus \Omega \subset \mathbb{R}^n \setminus B(x, \delta(x)),$$

and so

$$\int_\Omega \int_{\mathbb{R}^n \setminus \Omega} \frac{|u(x)|^p}{|x - y|^{n+sp}} \, dx \, dy \leq \int_\Omega \int_{\mathbb{R}^n \setminus B(x, \delta(x))} \frac{|u(x)|^p}{|x - y|^{n+sp}} \, dy \, dx$$

$$= \int_\Omega |u(x)|^p \left(n\omega_n \int_{\delta(x)}^\infty r^{-1-sp} dr \right) dx$$

$$= \frac{n\omega_n}{sp} \int_\Omega \frac{|u(x)|^p}{\delta(x)^{sp}} \, dx.$$

Suppose $ps > 1$. Then the fractional Hardy inequality (see [55], Theorem 1)

$$\int_\Omega \frac{|u(x)|^p}{\delta(x)^{sp}} \, dx \leq c \, [u]^p_{s,p,\Omega}$$

gives the result. When $ps < 1$ we use the inequality (see [55] and [44])

$$\int_\Omega \frac{|u(x)|^p}{\delta(x)^{sp}} \, dx \leq c \left([u]^p_{s,p,\Omega} + \|u\|^p_{p,\Omega} \right).$$

$\square$

In the borderline case $ps = 1$ it turns out that $\overset{0}{X}{}^s_p(\Omega) \neq \overset{0}{W}{}^s_p(\Omega)$: for details see [29], Remark 2.1, where it is shown that $\chi_\Omega \in \overset{0}{W}{}^s_p(\Omega) \setminus \overset{0}{X}{}^s_p(\Omega)$.

Consideration of nonlocal Dirichlet boundary conditions outside an open set $\Omega \subset \mathbb{R}^n$ makes it convenient to consider the homogeneous Sobolev–Slobodeckij space $\overset{0}{\mathcal{D}}{}^s_p(\Omega)$ defined by

$$\overset{0}{\mathcal{D}}{}^s_p(\Omega) = \text{completion of } C^\infty_0(\Omega) \text{ with respect to the norm } [\cdot]_{s,p,\mathbb{R}^n}.$$

The embedding $i \colon \overset{0}{\mathcal{D}}{}^s_p(\Omega) \to \overset{0}{\mathcal{D}}{}^s_p(\mathbb{R}^n)$ which associates to each $u \in \overset{0}{\mathcal{D}}{}^s_p(\Omega)$ its extension by 0 to all of $\mathbb{R}^n$ is well defined and continuous. If Ω supports the (s, p)-Friedrichs inequality, $\overset{0}{\mathcal{D}}{}^s_p(\Omega)$ is a space of functions continuously embedded in $L_p(\Omega)$; it then coincides with the closure in $W^s_p(\mathbb{R}^n)$ of $C^\infty_0(\Omega)$, namely $\overset{0}{X}{}^s_p(\Omega)$; and in view of the density of $C^\infty_0(\Omega)$ in $\overset{0}{\mathcal{D}}{}^s_p(\Omega)$ and Theorem 1.4.2.2 of [90] we see that if Ω supports the (s, p)-Friedrichs inequality, then

$$\overset{0}{\mathcal{D}}{}^s_p(\Omega) = \left\{ u \in W^s_p(\mathbb{R}^n) : u = 0 \text{ a.e. in } \mathbb{R}^n \backslash \Omega \right\}.$$

For the reader's convenience we now summarise in the next proposition the definitions and relationships between the various fractional spaces that have been introduced.

Proposition 3.7 *Let Ω be an open subset of $\mathbb{R}^n$, let $s \in (0, 1)$, suppose that $p \in (1, \infty)$ and let*

(i) $\overset{0}{W}{}^s_p(\Omega) :=$ *the closure of $C^\infty_0(\Omega)$ in $W^s_p(\Omega)$; equivalently, it is the completion of $C^\infty_0(\Omega)$ with respect to the norm $\|\cdot\|_{p,\Omega} + [\cdot]_{s,p,\Omega}$;*

(ii) $\overset{0}{X}{}^s_p(\Omega) :=$ *completion of $C^\infty_0(\Omega)$ with respect to the norm $\|\cdot\|_{p,\Omega} + [\cdot]_{s,p,\mathbb{R}^n}$;*

(iii) $\overset{0}{\mathcal{D}}{}^s_p(\Omega) :=$ *completion of $C^\infty_0(\Omega)$ with respect to the norm $[\cdot]_{s,p,\mathbb{R}^n}$.*

Then $\overset{0}{X}{}^s_p(\Omega) \subset \overset{0}{W}{}^s_p(\Omega)$; if Ω supports the (s, p)-Friedrichs inequality (and so, in particular, if Ω is bounded),

$$\overset{0}{X}{}^s_p(\Omega) = \overset{0}{\mathcal{D}}{}^s_p(\Omega) = \left\{ u \in W^s_p(\mathbb{R}^n) : u = 0 \text{ a.e. in } \mathbb{R}^n \backslash \Omega \right\}.$$

If Ω is bounded and has Lipschitz boundary, then if $sp \neq 1$,

$$\overset{0}{X}{}^s_p(\Omega) = \overset{0}{W}{}^s_p(\Omega);$$

while if $sp = 1$,

$$\overset{0}{X}{}^s_p(\Omega) \neq \overset{0}{W}{}^s_p(\Omega).$$

The uniform convexity and uniform smoothness of these spaces when $1 < p < \infty$ may be established much as in the case of $W^s_p(\Omega)$.

We next establish Hölder-continuity of the functions in fractional Sobolev spaces under certain conditions, just as for the classical spaces. The argument follows that given in Proposition 2.9 of [28].

Proposition 3.8 *Let $\Omega \subset \mathbb{R}^n$ be open and bounded, let $p \in (1, \infty)$, $s \in (0, 1)$ and suppose that $sp > n$. Then $\overset{0}{X}_p^s(\Omega) \hookrightarrow C^\alpha(\mathbb{R}^n)$, where $\alpha = s - n/p$: for all $u \in \overset{0}{X}_p^s(\Omega)$,*

$$|u(x) - u(y)| \le C(n, s, p) \left\| u | \overset{0}{X}_p^s(\Omega) \right\| |x - y|^\alpha \quad (x, y \in \mathbb{R}^n) \qquad (3.2.3)$$

and

$$\|u\|_{\infty, \mathbb{R}^n} \le C(n, s, p) \left\| u | \overset{0}{X}_p^s(\Omega) \right\| (\mathrm{diam}\,\Omega)^\alpha. \qquad (3.2.4)$$

Proof Let $u \in \overset{0}{X}_p^s(\Omega)$; by Proposition 3.7 we may regard u as an element of $\overset{0}{\mathcal{D}}_p^s(\Omega)$. Let $x_0 \in \mathbb{R}^n$ and $\delta > 0$; denote the mean value of u in $B(x_0, \delta)$ by $\bar{u}_{x_0, \delta}$. Then

$$\int_{B(x_0, \delta)} \left| u(x) - \bar{u}_{x_0, \delta} \right|^p dx \le \frac{1}{|B(x_0, \delta)|} \int_{B(x_0, \delta)} \int_{B(x_0, \delta)} |u(x) - u(y)|^p \, dx \, dy.$$

Since $|x - y| \le 2\delta$ for all $x, y \in B(x_0, \delta)$,

$$\int_{B(x_0, \delta)} \left| u(x) - \bar{u}_{x_0, \delta} \right|^p dx \le C \delta^{sp} [u]_{s, p, \mathbb{R}^n}^p,$$

and so

$$|B(x_0, \delta)|^{-sp/n} \int_{B(x_0, \delta)} \left| u(x) - \bar{u}_{x_0, \delta} \right|^p dx \le C' [u]_{s, p, \mathbb{R}^n}^p.$$

Hence u belongs to the Campanato space $\mathcal{L}^{p, sp}(\mathbb{R}^n)$, which is isomorphic to $C^\alpha(\mathbb{R}^n)$ with $\alpha = s - n/p$: see Section 1.3.2. From this, (3.2.3) follows. To obtain (3.2.4) take $y \in \mathbb{R}^n \backslash \mathrm{supp}\, u$. $\qquad\square$

A counterpart of the (s, p)-Friedrichs inequality, valid for all elements of $W_p^s(\Omega)$, is the (s, p)-*Poincaré inequality*. If $|\Omega| < \infty$, we say that Ω supports this inequality if there exists $K > 0$ such that for all $u \in W_p^s(\Omega)$,

$$\inf_{c \in \Phi} \|u - c\|_{p, \Omega} \le K [u]_{s, p, \Omega},$$

where Φ denotes the underlying field of scalars. Put

$$u_\Omega = |\Omega|^{-1} \int_\Omega u(x) \, dx$$

and observe that for all $c \in \Phi$,

$$\|u - u_\Omega\|_{p, \Omega} = \|u - c - (u - c)_\Omega\|_{p, \Omega} \le 2 \|u - c\|_{p, \Omega},$$

from which it follows that the Poincaré inequality may be equivalently written as

$$\|u - u_\Omega\|_{p,\Omega} \leq K_1 [u]_{s,p,\Omega}.$$

When Ω is bounded the following result holds.

Theorem 3.9 *Let $p \in [1, \infty)$, $s \in (0, 1)$ and suppose that Ω is bounded. Then for all $u \in W_p^s(\Omega)$,*

$$\|u - u_\Omega\|_{p,\Omega}^p \leq \frac{(\text{diam } \Omega)^{n+sp}}{|\Omega|} \int_\Omega \int_\Omega \frac{|u(y) - u(x)|^p}{|y - x|^{n+sp}} \, dy \, dx,$$

where $u_\Omega = |\Omega|^{-1} \int_\Omega u(x) \, dx$.

Proof By Jensen's inequality (see, for example, [97], p. 202),

$$
\begin{aligned}
\|u - u_\Omega\|_{p,\Omega}^p &= \int_\Omega \left| |\Omega|^{-1} \int_\Omega (u(y) - u(x)) \, dx \right|^p dy \\
&\leq |\Omega|^{-1} \int_\Omega \int_\Omega |u(y) - u(x)|^p \, dx \, dy \\
&\leq \frac{(\text{diam } \Omega)^{n+sp}}{|\Omega|} \cdot \int_\Omega \int_\Omega \frac{|u(y) - u(x)|^p}{|y - x|^{n+sp}} \, dy \, dx.
\end{aligned}
$$

$\square$

Remark 3.10

(i) This result underlines the difference between fractional Sobolev spaces and their classical counterparts: the (s, p)-Poincaré inequality in $W_p^s(\Omega)$ holds for all (bounded) Ω, while the classical Poincaré inequality in $W_p^1(\Omega)$ does not hold in all bounded sets Ω: see [64], Theorem V.4.21(iii). The validity of the fractional case does exhibit some domain dependence if the stronger inequality

$$\inf_{c \in \mathbb{R}} \|u - c\|_{p,\Omega}^p \leq C \int_\Omega \int_{\Omega \cap B(x, \tau \, dist(x, \partial\Omega))} \frac{|u(y) - u(x)|^p}{|y - x|^{n+sp}} \, dy \, dx$$

(with $\tau \in (0, 1)$) is considered. In [52] it is shown that this may fail in certain β−John domains (see [52] and [99] for the definition of these domains), while in [99] the inequality is established for 1−John domains. The paper [52] also discusses the influence of weights. Moreover, the double integral on the right-hand side of the displayed formula is comparable to the full Gagliardo seminorm under suitable conditions on Ω, such as being Lipschitz [55] or uniform [149].

Definitions of plumpness different from that given above and used by Zhou may be found in the literature. Sets that are plump in his sense are often called *lower Ahlfors regular*.

(ii) Note that when Ω is a cube Q, the theorem gives

$$\left\| u - u_Q \right\|_{p,Q} \leq n^{(n+sp)/(2p)} |Q|^{s/n} [u]_{s,p,\Omega} \, .$$

The constant in this inequality is not sharp: see Theorem 1 of [23].

When $s \in \mathbb{N}$ and $p \in [1, \infty)$, elements of $W_p^s(\Omega)$ may be extended to be functions in $W_p^s(\mathbb{R}^n)$ if the domain Ω has some regularity properties: we have already used such a result in proving Proposition 3.1 (ii). Here we describe work by Zhou [173] that characterises those Ω for which corresponding extension results hold for arbitrary $s \in (0, 1)$.

Definition 3.11   Let Ω be a domain in $\mathbb{R}^n$ ($n \geq 2$) and suppose that $p \in [1, \infty)$ and $s > 0$. The set Ω is called a W_p^s extension domain if given any $u \in W_p^s(\Omega)$ there exists $\tilde{u} \in W_p^s(\mathbb{R}^n)$ such that $\tilde{u}|_\Omega = u$ and

$$\left\| \tilde{u} | W_p^s(\mathbb{R}^n) \right\| \leq C \left\| u | W_p^s(\Omega) \right\| ,$$

where C is a constant that depends on n, p, s, Ω but not on u. It is said to be plump if there is a constant $c > 0$ such that for all $x \in \Omega$ and all $r \in (0, 1]$,

$$|B(x, r) \cap \Omega| \geq cr^n .$$

When $s \in \mathbb{N}$ every Ω with minimally smooth boundary (see [64], V.4) is a W_p^s extension domain; in particular, every bounded Ω with boundary of class $C^{0,1}$ is an extension domain. Zhou's result implies the following.

Theorem 3.12   *Let Ω be a domain in $\mathbb{R}^n$ ($n \geq 2$). Then Ω is a W_p^s extension domain for all $s \in (0, 1)$ and all $p \in [1, \infty)$ if and only if it is plump.*

Remark 3.13

1. When $s > 0$, $s \notin \mathbb{N}$ and $p \in [1, \infty)$, the space $W_p^s(\mathbb{R}^n)$ defined above coincides with the Besov space $B_{p,p}^s(\mathbb{R}^n)$, which itself coincides with the Lizorkin–Triebel space $F_{p,p}^s(\mathbb{R}^n)$: see [169], 2.3–2.5 and [95], 3.6. This means that all the embedding theorems known for $B_{p,p}^s(\mathbb{R}^n)$ (see, for example, [70], 2.3.3) are also available for $W_p^s(\mathbb{R}^n)$. When Ω is an open subset of $\mathbb{R}^n$ the spaces $B_{p,p}^s(\Omega)$ (see, for example, the Proposition in [169], 3.28) may be defined intrinsically or by restriction of elements of $B_{p,p}^s(\mathbb{R}^n)$: if $\partial\Omega$ is smooth enough the resulting space is the same whichever procedure is adopted, and so $W_p^s(\Omega)$ may be identified with $B_{p,p}^s(\Omega)$ (see, for example, [169]). Thus embedding results known for $B_{p,p}^s(\Omega)$ (see, for example, [70], 2.5) hold for $W_p^s(\Omega)$. Hence Proposition 3.1 (ii) could have been deduced from the known embeddings for Besov spaces: we chose to give an independent proof so as to acclimatise the reader to the techniques to be used later on. To illustrate what may be obtained from results known for Besov spaces, we give the following:

1. Let Ω be a bounded open subset of $\mathbb{R}^n$ with C^∞ boundary, let $s_1, s_2 \in (0, \infty) \backslash \mathbb{N}$ and $p_1, p_2 \in (1, \infty)$. Then if $s_1 - s_2 - n \max (1/p_1 - 1/p_2, 0) \geq 0$,

$$W_{p_1}^{s_1}(\Omega) \hookrightarrow W_{p_2}^{s_2}(\Omega);$$

the embedding is compact if $s_1 - s_2 - n \max (1/p_1 - 1/p_2, 0) > 0$. Similar statements hold for the spaces $\overset{0}{W}{}_{p_i}^{s_i}(\Omega)$; and if $s_i p_i \neq 1$ $(i = 1, 2)$, each $\overset{0}{W}{}_{p_i}^{s_i}(\Omega)$ can be replaced by $\overset{0}{X}{}_{p_i}^{s_i}(\Omega)$ (see Proposition 3.7).

2. Given $u \in \mathcal{S}(\mathbb{R}^n)$, its restriction to the hyperplane

$$\Gamma := \left\{ x = (x', x_n) \in \mathbb{R}^n : x_n = 0 \right\},$$

that is, the function $v \in \mathcal{S}(\mathbb{R}^{n-1})$ given by $v(x') = u(x', 0)$ $(x' \in \mathbb{R}^{n-1})$, is called the trace of u on Γ. It turns out that, given any $s > 1/2$, the map $u \longmapsto v$ can be extended to a continuous, surjective map $tr_\Gamma \colon W_2^s(\mathbb{R}^n) \to W_2^{s-1/2}(\mathbb{R}^{n-1})$, the trace map. For a proof of this assertion, its extension to traces on the boundary of bounded, smoothly bounded subsets of $\mathbb{R}^n$ and in the context of more general function spaces, we refer to [95], Chapter 4 and [142], Section 3.

3. If $p \in [1, \infty)$ and $s \in (0, 1)$, then the real space $W_p^s(\mathbb{R}^n)$ has the truncation property: if u belongs to $W_p^s(\mathbb{R}^n)$ then so does $u_+ := \max(u, 0)$ and

$$\left\| u_+ | W_p^s(\mathbb{R}^n) \right\| \leq c \left\| u | W_p^s(\mathbb{R}^n) \right\|,$$

where c is a constant independent of u. For this, and many more general results, we refer to [171], Section 25. $\qquad\square$

We have seen earlier that when Ω is bounded and $1 \leq q < p < \infty$, the space $W_p^s(\Omega)$ is compactly embedded in $L_q(\Omega)$, no conditions on $\partial\Omega$ being necessary. When $q = p$ this embedding is not always compact. On the positive side, when Ω has smooth enough boundary, $W_p^s(\Omega)$ can be identified with a Besov space and the compactness of $I \colon W_p^s(\Omega) \to L_p(\Omega)$ follows from known properties of these spaces. The same holds when Ω is an extension domain: see Theorem 7.1 of [142], which also contains an example showing that some condition on $\partial\Omega$ is needed for I to be compact. To investigate the position a little more closely we introduce the notions of entropy and approximation numbers, and the measure of noncompactness of a map, following the line of argument for the non-fractional case given in [64], V.5. where background material may be found.

Given a bounded linear map T between Banach spaces X and Y, for each $k \in \mathbb{N}$ the kth *entropy number* $e_k(T)$ of T is defined by

$$e_k(T) = \inf \left\{ \varepsilon > 0 : T(B_X) \text{ can be covered by } 2^{k-1} \text{ balls in } Y \text{ with radius } \varepsilon \right\},$$

where B_X is the closed unit ball in X. Since T is compact if and only if $\lim_{k \to \infty} e_k(T) = 0$, this limit is called the *measure of noncompactness* of T; we denote it by $\beta(T)$. Evidently $0 \le \beta(T) \le \|T\|$; if $\beta(T) = \|T\|$ we say that T is *maximally noncompact*: for examples of such maps and further details see [69] and the references given in that paper. In fact, [69] shows that when $p \in (1, \infty)$ and Ω is an infinite strip, the natural embedding $I_p \colon W_p^1(\Omega) \to L_p(\Omega)$ is maximally noncompact in the sense that $\beta(I_p) = \|I_p\|$; from [68] it turns out that I_p is not strictly singular. For each $k \in \mathbb{N}$ the kth *approximation number* $a_k(T)$ of T is defined by

$$a_k(T) = \inf \{ \|T - F\| : F \in B(X, Y), \ \mathrm{rank}\, F < k \};$$

$\lim_{k \to \infty} a_k(T)$ is denoted by $\alpha(T)$. If $\alpha(T) = 0$ the map T is compact; the converse is true if Y has the approximation property. By Proposition II.2.7 of [64],

$$\beta(T) \le \alpha(T).$$

We now consider the embedding map $I \colon W_p^s(\Omega) \to L_p(\Omega)$, where $|\Omega| < \infty$, $s \in (0, 1)$ and $p \in (1, \infty)$. First suppose that Ω is bounded. Given $\varepsilon > 0$, write $\Omega_\varepsilon = \{ x \in \Omega : d(x, \partial\Omega) < \varepsilon \}$; by Theorem V.4.20 of [64] there is a domain U_ε with analytic boundary such that $\Omega \backslash \Omega_\varepsilon \subset U_\varepsilon \subset \overline{U_\varepsilon} \subset \Omega$; put $\mathcal{U} = \cup_{\varepsilon > 0} U_\varepsilon$. For each $\varepsilon > 0$, the natural embedding $I_\varepsilon \colon W_p^s(\Omega) \to L_p(U_\varepsilon)$ is compact since it can be represented as the composition of the maps

$$W_p^s(\Omega) \hookrightarrow W_p^s(U_\varepsilon) \hookrightarrow\hookrightarrow L_p(U_\varepsilon).$$

Let $F(\Omega)$ (resp. $F(U_\varepsilon)$) stand for the family of all bounded linear maps from $W_p^s(\Omega)$ to $L_p(\Omega)$ (resp. $L_p(U_\varepsilon)$) that have finite rank. As $L_p(U_\varepsilon)$ has the approximation property, it follows from Theorem 1.2.25 of [61] that there exists $P \in F(U_\varepsilon)$ such that for all $f \in W_p^s(\Omega)$,

$$\|f - Pf\|_{p, U_\varepsilon} \le \varepsilon \|f\|_{s, p, \Omega}.$$

Since $L_p(\Omega)$ has the approximation property,

$$\begin{aligned}
\alpha(I) &= \mathrm{dist}\left(I, K\left(W_p^s(\Omega), L_p(\Omega)\right)\right) \\
&= \inf \left\{ \|I - K\| : K \in K\left(W_p^s(\Omega), L_p(\Omega)\right) \right\}.
\end{aligned}$$

In fact, $\beta(I) = \alpha(I)$. To prove this we first establish a lemma.

Lemma 3.14 *Let $P \in F(\Omega)$. Then given $\varepsilon > 0$, there exists $R \in F(\Omega)$ and a domain $\Omega' \subset\subset \Omega$ such that $\|P - R\| < \varepsilon$ and $R\left(W_p^s(\Omega)\right) \subset C_0^\infty(\Omega')$. If Ω_0 ($\subset \Omega$) is open, $\varepsilon > 0$ and $P \in F(\Omega)$ are given, then there exists $R \in F(\Omega)$ such that $\|(P - R)f\|_{p, \Omega_0} \le \varepsilon \|f\|_{s, p, \Omega}$ for all $f \in W_p^s(\Omega)$ and $R\left(W_p^s(\Omega)\right) \subset C_0^\infty(\Omega_0)$.*

Proof There are linearly independent functions $u_1, ..., u_N$, with each $\|u_i\|_{p,\Omega} = 1$, such that for each $f \in W_p^s(\Omega)$, $Pf = \sum_{i=1}^{N} c_i(f) u_i$. Since all norms on the finite-dimensional range of P are equivalent, there exists K such that

$$\sum_{i=1}^{N} |c_i(f)| \le K \|Pf\|_{p,\Omega} \le K \|P\| \|f\|_{s,p,\Omega}.$$

For each i let $\phi_i \in C_0^\infty(\Omega)$ be such that $\|u_i - \phi_i\|_{p,\Omega} < \varepsilon / (K \|P\|)$ and set $Rf = \sum_{i=1}^{N} c_i(f) \phi_i$. Then $R \in F(\Omega)$ and

$$\|Pf - Rf\|_{p,\Omega} \le \sum_{i=1}^{N} |c_i(f)| \|u_i - \phi_i\|_{p,\Omega} \le \varepsilon \|f\|_{s,p,\Omega};$$

since supp $Rf \subset \cup_{i=1}^{N}$ supp $\phi_i \subset\subset \Omega$, the first part of the lemma follows. The proof of the second part is similar, noting that $f \longmapsto \chi_{\Omega_0} Pf \in F(\Omega_0)$ and choosing $\phi_i \in C_0^\infty(\Omega_0)$. $\qquad\square$

Given any $\varepsilon > 0$, let

$$\Gamma(\varepsilon) := \sup \left\{ \|u\|_{p,\Omega_\varepsilon}^p : \|u\|_{s,p,\Omega} = 1 \right\},$$

and

$$\Gamma(0) := \lim_{\varepsilon \to 0} \Gamma(\varepsilon).$$

Theorem 3.15 *When Ω is bounded, $\Gamma(0) = \beta(I)^p = \alpha(I)^p$.*

Proof This is exactly the same as that of Theorem V.5.7 of [64]. It is in this proof that the material involving $\mathcal{U} = \cup_{\varepsilon > 0} U_\varepsilon$ is used. $\qquad\square$

The next theorem shows how the universal validity of the fractional Poincaré inequality (in bounded sets Ω) affects the measure of noncompactness $\beta(I)$.

Theorem 3.16 *Suppose that Ω is bounded. Then $\beta(I) = \alpha(I) < 1$.*

Proof The map $f \longmapsto f_\Omega$ belongs to $F(\Omega)$, and so, by Lemma 3.14, given $\varepsilon > 0$, there exist $R \in F(\Omega)$ and $\delta > 0$ such that for all $f \in W_p^s(\Omega)$,

$$\|f_\Omega - Rf\|_{p,\Omega}^p \le 2^{1-p} \varepsilon \|f\|_{s,p,\Omega}^p$$

and supp $Rf \subset \Omega \setminus \Omega^\delta := \{x \in \Omega : d(x, \partial\Omega) > \delta\}$. Hence by the Poincaré inequality,

$$\|f - Rf\|_{p,\Omega}^p \le 2^{p-1} K^p [f]_{s,p,\Omega}^p + \varepsilon \|f\|_{s,p,\Omega}^p$$
$$\le (K_1 + \varepsilon) [f]_{s,p,\Omega}^p + \varepsilon \|f\|_{p,\Omega}^p,$$

where $K_1 = 2^{p-1}K$. Thus

$$\|f\|_{p,\Omega\setminus\Omega^\delta}^p \leq (K_1 + \varepsilon)\,[f]_{s,p,\Omega}^p + \varepsilon\,\|f\|_{p,\Omega}^p$$
$$= (K_1 + \varepsilon)\left(\|f\|_{s,p,\Omega}^p - \|f\|_{p,\Omega}^p\right) + \varepsilon\,\|f\|_{p,\Omega}^p$$
$$\leq (K_1 + \varepsilon)\,\|f\|_{s,p,\Omega}^p - K_1\,\|f\|_{p,\Omega^\delta}^p,$$

and so

$$\|f\|_{p,\Omega\setminus\Omega^\delta}^p \leq \left[(K_1 + \varepsilon)/(K_1 + 1)\right]\|f\|_{s,p,\Omega}^p.$$

Thus by Theorem 3.15, $\alpha(I) < 1$. □

This result shows that $\beta(I)^p \leq 1 - 1/(K_1 + 1)$, where

$$K_1 = 2^p\,(\text{diam }\Omega)^{n+sp}\,/\,|\Omega| = 2^p L(\Omega)\,(\text{diam }\Omega)^{sp},$$

where

$$L(\Omega) = (\text{diam }\Omega)^n\,/\,|\Omega| \geq 1.$$

Some idea of the dependence of $\beta(I)$ on s may be obtained from this. For example, it can be shown that

$$\beta(I)^p \leq 1 - \frac{1}{2\max\{1,\,2^p L(\Omega)\,(\text{diam }\Omega)^{sp}\}}.$$

In particular, if $(\text{diam }\Omega)^{sp} > 2^{-p}/L(\Omega)$, then

$$\beta(I)^p \leq 1 - \frac{(\text{diam }\Omega)^{-sp}}{2^{p+1}L(\Omega)}.$$

Thus if Ω is the unit cube $(0,1)^n$, so that $|\Omega| = 1$ and diam $\Omega = \sqrt{n}$,

$$\beta(I)^p \leq 1 - 2^{-p-1}n^{-(sp+n)/2}.$$

Now suppose that the open set Ω is merely required to have finite measure. We aim to obtain results analogous to those given in [64], V.5.3 for the classical Sobolev case that involve the map $f \mapsto \nabla f$ of $W_p^1(\Omega)$ to $\left[L_p(\Omega)\right]^n$. Define $T: W_p^s(\Omega) \to L_p(\Omega \times \Omega)$ by

$$(Tf)(x,y) = \frac{f(x) - f(y)}{|x-y|^{\frac{n}{p}+s}}\,((x,y) \in \Omega \times \Omega).$$

The reduced minimum modulus $\gamma(T)$ of T is defined by

$$\gamma(T) = \inf\left\{\|Tf\|_{p,\Omega\times\Omega}\,/\text{dist}\,(f, \ker T) : f \in W_p^s(\Omega)\setminus\{0\}\right\}.$$

Note that ker T can be identified with Φ, the underlying field of scalars, and

$$\text{dist}\,(f, \ker T) = \inf_\Phi \|f - c\|_{s,p,\Omega}.$$

We recall that the (s, p)-Poincaré inequality asserts that there exists $K > 0$ such that for all $f \in W_p^s(\Omega)$,

$$\inf_{c \in \mathbb{C}} \|f - c\|_{p,\Omega} \leq K [f]_{s,p,\Omega} \,,$$

and that if Ω is bounded, this form of the Poincaré inequality holds with no additional assumption on Ω.

Assume that the (s, p)-Poincaré inequality holds. Then for all $f \in W_p^s(\Omega) \backslash \{0\}$ and all $c \in \Phi$,

$$\|Tf\|_{p,\Omega \times \Omega} / \text{dist } (f, \ker T) \geq [f]_{s,p,\Omega} / \|f - c\|_{s,p,\Omega}$$

$$= \frac{[f]_{s,p,\Omega}}{\left[\|f - c\|_{p,\Omega}^p + [f]_{s,p,\Omega}^p \right]^{1/p}}$$

$$\geq \frac{[f]_{s,p,\Omega}}{\left[K^p [f]_{s,p,\Omega}^p + [f]_{s,p,\Omega}^p \right]^{1/p}}$$

$$= (1 + K^p)^{-1/p} \,.$$

Thus $\gamma(T) > 0$ and so T has closed range, by Theorem 1.3.4 of [64]. Conversely, if $\gamma(T) > 0$, then for all $f \in W_p^s(\Omega)$,

$$\inf_{c \in \mathbb{C}} \|f - c\|_{s,p,\Omega} \leq \gamma(T)^{-1} \|Tf\|_{p,\Omega \times \Omega} \,,$$

so that the Poincaré inequality holds and T has closed range.

To summarise the position:

(i) if $|\Omega| < \infty$, then the Poincaré inequality holds if and only if T has closed range;
and

(ii) if Ω is bounded, then the Poincaré inequality holds, T has closed range and for the embedding $I \colon W_p^s(\Omega) \to L_p(\Omega)$ we have

$$\alpha(I) = \beta(I) < 1;$$

more precisely (see Theorem 3.16),

$$\alpha^p(I) = \beta^p(I) \leq 1 - 1/(C + 1),$$

where

$$C = 2^p (\text{diam } \Omega)^{n+sp} / |\Omega| \,.$$

A similar analysis may be carried out for the Friedrichs inequality and its connection with the embedding of $\overset{0}{X}_p^s(\Omega)$ in $L_p(\Omega)$. $\qquad\square$

Next we give a fractional analogue of Proposition 2.6 in the form given by Lemma A.1 of [28], dealing with the behaviour of functions under translation.

Proposition 3.17 *Let $p \in [1, \infty)$ and $s \in (0, 1)$. Then there is a constant $C = C(n, p)$ such that for every $u \in C_0^\infty(\mathbb{R}^n)$,*

$$\sup_{|h|>0} \int_{\mathbb{R}^n} \frac{|u(x+h) - u(x)|^p}{|h|^{sp}} dx \le C(1-s) [u]_{s,p,\mathbb{R}^n}^p.$$

Proof Let $\rho \in C_0^\infty(\mathbb{R}^n)$ be non-negative, with supp $\rho \subset B(0, 1) \backslash B(0, 1/2)$ and $\int_{\mathbb{R}^n} \rho dx = 1$. Given $h \in \mathbb{R}^n \backslash \{0\}$, put

$$\rho_\varepsilon(x) = \varepsilon^{-n} \rho(x/\varepsilon) \ (x \in \mathbb{R}^n, 0 < \varepsilon < |h|).$$

Using suitable changes of variable we see that for all $x \in \mathbb{R}^n$,

$$
\begin{aligned}
|u(x+h) - u(x)| &= \left| \int_{\mathbb{R}^n} (u(x+h) - u(x)) \rho_\varepsilon(y) \, dy \right| \\
&= \left| \int_{\mathbb{R}^n} \left\{ \begin{aligned} &[u(x+h) - u(x+h-y)] + u(x+h-y) \\ &- [u(x) - u(x-y)] - u(x-y) \end{aligned} \right\} \rho_\varepsilon(y) \, dy \right| \\
&\le \left| \int_{\mathbb{R}^n} u(y) \left[\rho_\varepsilon(x+h-y) - \rho_\varepsilon(x-y) \right] dy \right| \\
&\quad + \int_{\mathbb{R}^n} |u(x+h) - u(x+h-y)| \rho_\varepsilon(y) \, dy \\
&\quad + \int_{\mathbb{R}^n} |u(x) - u(x-y)| \rho_\varepsilon(y) \, dy.
\end{aligned}
$$

Since $\int_{\mathbb{R}^n} \nabla \rho_\varepsilon \, dx = 0$ it follows that

$$
\begin{aligned}
&\left| \int_{\mathbb{R}^n} u(y) \left[\rho_\varepsilon(x+h-y) - \rho_\varepsilon(x-y) \right] dy \right| \\
&= \left| \int_0^1 \int_{\mathbb{R}^n} u(y) \langle \nabla \rho_\varepsilon(x-y+sh), h \rangle \, dy \, ds \right|
\end{aligned}
$$

may be written as

$$
\begin{aligned}
&\left| \int_0^1 \int_{\mathbb{R}^n} [u(y) - u(x+sh)] \langle \nabla \rho_\varepsilon(x-y+sh), h \rangle \, dy \, ds \right| \\
&\le \|\nabla \rho\|_\infty |h| \varepsilon^{-n-1} \int_0^1 \int_{B(x+sh,\varepsilon)\backslash B(x+sh,\varepsilon/2)\backslash} |u(y) - u(x+sh)| \, dy \, ds \\
&= \|\nabla \rho\|_\infty |h| \varepsilon^{-n-1} \int_0^1 \int_{B(0,\varepsilon)\backslash B(0,\varepsilon/2)} |u(x+z+sh) - u(x+sh)| \, dz \, ds.
\end{aligned}
$$

Use of Jensen's inequality (see, for example, [126], Theorem 2.2) together with the translation invariance of the L_p norm now shows that

$$\int_{\mathbb{R}^n} |u(x+h) - u(x)|^p \, dx$$

is bounded above by

$$C \, |h|^p \, \varepsilon^{-n-p} \int_{B(0,\varepsilon) \setminus B(0,\varepsilon/2)} \int_{\mathbb{R}^n} |u(x+z) - u(x)|^p \, dx \, dz$$

$$+ C \, \|\rho\|_\infty \, \varepsilon^{-n} \int_{B(0,\varepsilon) \setminus B(0,\varepsilon/2)} \int_{\mathbb{R}^n} |u(x+z) - u(x)|^p \, dx \, dz.$$

As $\varepsilon < |h|$ we obtain

$$\int_{\mathbb{R}^n} |u(x+h) - u(x)|^p \, dx$$

$$\leq C_1 \, |h|^p \, \varepsilon^{-n-p} \int_{B(0,\varepsilon) \setminus B(0,\varepsilon/2)} \int_{\mathbb{R}^n} |u(x+z) - u(x)|^p \, dx \, dz.$$

Hence

$$\int_{\mathbb{R}^n} \frac{|u(x+h) - u(x)|^p}{|h|^{sp}} \, dx$$

$$\leq C_1 \, |h|^{p(1-s)} \, \varepsilon^{-n-p} \int_{B(0,\varepsilon) \setminus B(0,\varepsilon/2)} \int_{\mathbb{R}^n} |u(x+z) - u(x)|^p \, dx \, dz.$$

Multiply both sides by $\varepsilon^{p(1-s)-1}$, integrate with respect to ε from 0 to $|h|$ and simplify: we find that

$$\frac{1}{p(1-s)} \int_{\mathbb{R}^n} \frac{|u(x+h) - u(x)|^p}{|h|^{sp}} \, dx$$

is bounded above by

$$C_1 \int_0^{|h|} \varepsilon^{-n-ps-1} \int_{B(0,\varepsilon)} \int_{\mathbb{R}^n} |u(x+z) - u(x)|^p \, dx \, dz \, d\varepsilon.$$

Next put

$$G(\varepsilon) = \int_{B(0,\varepsilon)} \int_{\mathbb{R}^n} |u(x+z) - u(x)|^p \, dx \, dz, \, 0 < \varepsilon < |h| \, .$$

Observe that $G(0) = 0$ and G is increasing. For small positive τ,

$$\int_0^{|h|} \frac{(G(t) - \tau)_+}{t^{n+ps+1}} \, dt = \frac{-1}{n+ps} \cdot \frac{(G(|h|) - \tau)_+}{|h|^{n+ps}} + \frac{1}{n+ps} \int_{\{G(t)>\tau\}} \frac{G'(t)}{t^{n+ps}} \, dt$$

$$\leq \frac{1}{n+ps} \int_0^{|h|} \frac{G'(t)}{t^{n+ps}} \, dt.$$

Letting $\tau \to 0$ we see that

$$(n + ps) \int_0^{|h|} t^{-n-ps-1} G(t)\, dt \leq \int_0^{|h|} t^{-n-ps} G'(t)\, dt$$

which equals

$$\int_0^{|h|} t^{-n-ps} \int_{\partial B(0,t)} \int_{\mathbb{R}^n} |u(x+z) - u(x)|^p \, dx \, d\sigma(z)\, dt$$
$$= \int_{B(0,|h|)} \int_{\mathbb{R}^n} \frac{|u(x+z) - u(x)|^p}{|z|^{n+ps}} dx\, dz \leq [u]_{s,p,\mathbb{R}^n}^p ,$$

from which the result follows. $\qquad\qquad\qquad\qquad\qquad\qquad\qquad\square$

We now turn to the limiting behaviour of the fractional spaces: more precisely, to their connection with the classical Sobolev spaces as the parameter s approaches 1 or 0. To do this we introduce a family $(\rho_\varepsilon)_{\varepsilon>0}$ of non-negative functions, each belonging to $L_{1,loc}(0,\infty)$, such that

$$\int_0^\infty \rho_\varepsilon(r) r^{n-1} dr = 1 \ (\varepsilon > 0), \qquad\qquad (3.2.5)$$

and

$$\lim_{\varepsilon \to 0} \int_\delta^\infty \rho_\varepsilon(r) r^{n-1} dr = 0 \text{ for all } \delta > 0. \qquad\qquad (3.2.6)$$

Note that for such a family,

$$\lim_{\varepsilon \to 0} \int_0^R \rho_\varepsilon(r) r^{p+n-1} dr = 0 \ (0 < R < \infty, 1 \leq p < \infty). \qquad (3.2.7)$$

It is plainly enough to establish this for $R \in (1, \infty)$. Then for all $\delta \in (0, 1)$,

$$\int_0^R \rho_\varepsilon(r) r^{p+n-1} dr = \int_0^\delta \rho_\varepsilon(r) r^{p+n-1} dr + \int_\delta^R \rho_\varepsilon(r) r^{p+n-1} dr$$
$$\leq \delta^p + R^p \int_\delta^R \rho_\varepsilon(r) r^{n-1} dr,$$

so that

$$\limsup_{\varepsilon \to 0} \int_0^R \rho_\varepsilon(r) r^{p+n-1} dr \leq \delta^p.$$

Since this is true for all $\delta \in (0, 1)$, the claim follows.

Lemma 3.18 *Let $p \in (1, \infty)$, suppose that $f \in W_p^1(\mathbb{R}^n)$ and let $\rho \in L_1(\mathbb{R}^n)$, $\rho \geq 0$. Then*

$$\int_{\mathbb{R}^n} \int_{\mathbb{R}^n} \frac{|f(x) - f(y)|^p}{|x - y|^p} \rho(x - y)\, dx\, dy \leq \|\rho\|_1 \int_{\mathbb{R}^n} |\nabla f|^p\, dx. \qquad (3.2.8)$$

Proof By Proposition 2.6,

$$\left(\int_{\mathbb{R}^n} |f(x+h) - f(x)|^p \, dx \right)^{1/p} \leq |h| \, \||\nabla f|\|_p$$

for all $h \in \mathbb{R}^n$. Hence

$$\int_{\mathbb{R}^n} \int_{\mathbb{R}^n} \frac{|f(x) - f(y)|^p}{|x-y|^p} \rho\,(x-y)\,dx\,dy = \int_{\mathbb{R}^n} \frac{\rho(h)}{|h|^p} \int_{\mathbb{R}^n} |f(x+h) - f(x)|^p \, dx \, dh$$

$$\leq \|\rho\|_1 \int_{\mathbb{R}^n} |\nabla f|^p \, dx.$$

$\square$

Theorem 3.19 *Suppose that $f \in L_p\,(\mathbb{R}^n)$ $(1 < p < \infty)$. Then, with ρ_ε defined as above,*

$$\lim_{\varepsilon \to 0} \int_{\mathbb{R}^n} \int_{\mathbb{R}^n} \frac{|f(x) - f(y)|^p}{|x-y|^p} \rho_\varepsilon\,(|x-y|)\,dx\,dy = K(p,n) \int_{\mathbb{R}^n} |\nabla f|^p \, dx, \quad (3.2.9)$$

where

$$K(p,n) = \frac{\Gamma\left(\frac{p+1}{2}\right) \Gamma\left(\frac{n}{2}\right)}{\sqrt{\pi}\,\Gamma\left(\frac{n+p}{2}\right)},$$

with the understanding that $\int_{\mathbb{R}^n} |\nabla f|^p \, dx = \infty$ if $f \notin W_p^1\,(\mathbb{R}^n)$.

Proof Let

$$F_\varepsilon(x,y) = \frac{|f(x) - f(y)|}{|x-y|} \rho_\varepsilon^{1/p}\,(|x-y|)\,(x, y \in \mathbb{R}^n, x \neq y, \varepsilon > 0),$$

and suppose first that $f \in W_p^1\,(\mathbb{R}^n)$. We have to show that

$$\lim_{\varepsilon \to 0} \|F_\varepsilon\|_{p, \mathbb{R}^n \times \mathbb{R}^n}^p = K \, \||\nabla f|\|_p^p \quad (3.2.10)$$

with $K = K(p,n)$. By Lemma 3.18, for all $\varepsilon > 0$ and all $g \in W_p^1\,(\mathbb{R}^n)$,

$$\left| \|F_\varepsilon\|_p - \|G_\varepsilon\|_p \right| \leq \|F_\varepsilon - G_\varepsilon\|_p \leq \omega_{n-1} \, \||\nabla(f - g)|\|_p$$

where G_ε is defined in the same way as F_ε. It is therefore enough to prove (3.2.10) when f belongs to any dense subset of $W_p^1\,(\mathbb{R}^n)$. Thus we assume that $f \in C_0^2\,(\mathbb{R}^n)$, with supp f contained in a bounded open set $\Omega \subset \mathbb{R}^n$; for each fixed $x \in \Omega$ let $R = \text{dist}\,(x, \partial\Omega)$; note that

$$\frac{|f(x) - f(y)|}{|x-y|} = \left| (\nabla f)\,(x) \cdot \frac{x-y}{|x-y|} \right| + O(|x-y|).$$

Then

$$\int_{\mathbb{R}^n} \frac{|f(x) - f(y)|^p}{|x-y|^p} \rho_\varepsilon\,(|x-y|)\,dy = \int_{B(x,R)} + \int_{\mathbb{R}^n \setminus B(x,R)} := I_1(x) + I_2(x).$$

In view of (3.2.6),

$$I_2(x) \le \frac{|f(x)|^p}{R^p} \int_{\mathbb{R}^n \setminus B(x,R)} \rho_\varepsilon\left(|x-y|\right) dy \to 0 \text{ as } \varepsilon \to 0.$$

Moreover,

$$I_1(x) = \int_0^R \rho_\varepsilon(r) \int_{|y-x|=r} \left(\left| (\nabla f)(x) \cdot \frac{x-y}{|x-y|} \right|^p + O\left(|x-y|^p\right) \right) d\sigma\, dr$$

$$= \int_0^R \rho_\varepsilon(r) \int_{|\omega|=r} \left(\left| (\nabla f)(x) \cdot \frac{\omega}{|\omega|} \right|^p + O\left(r^p\right) \right) d\sigma\, dr$$

$$= |(\nabla f)(x)|^p \left(\int_{S^{n-1}} |\omega \cdot \mathbf{e}|^p\, d\sigma \right) \int_0^R r^{n-1} \rho_\varepsilon(r)\, dr$$

$$+ O\left(\int_0^R r^{n+p-1} \rho_\varepsilon(r)\, dr \right)$$

where $\mathbf{e}$ is a unit vector in $\mathbb{R}^n$: by (3.3.7) of [15],

$$\int_{S^{n-1}} |\omega \cdot \mathbf{e}|^p\, d\sigma = K(p,n).$$

Hence

$$\lim_{\varepsilon \to 0} \int_{\mathbb{R}^n} \frac{|f(x)-f(y)|^p}{|x-y|^p} \rho_\varepsilon\left(|x-y|\right) dy = K(p,n) \left|(\nabla f)(x)\right|^p \text{ for all } x \in \Omega.$$

(3.2.11)

Since $f \in C_0^2(\mathbb{R}^n)$, there exists M such that $|f(x)-f(y)| \le M|x-y|$ for all $x, y \in \Omega$: thus

$$\int_{\mathbb{R}^n} \frac{|f(x)-f(y)|^p}{|x-y|^p} \rho_\varepsilon\left(|x-y|\right) dy \le M^p \omega_{n-1} \quad (x \in \Omega, k \in \mathbb{N}).$$

Together with (3.2.11) and the dominated convergence theorem this establishes (3.2.10) for all $f \in C_0^2(\mathbb{R}^n)$ and hence for all $f \in W_p^1(\mathbb{R}^n)$.

To complete the proof it will be enough to show that if $f \in L_p(\mathbb{R}^n)$ and

$$A_p := \liminf_{\varepsilon \to 0} \left(\int_{\mathbb{R}^n} \int_{\mathbb{R}^n} \frac{|f(x)-f(y)|^p}{|x-y|^p} \rho_\varepsilon\left(|x-y|\right) dx\, dy \right)^{1/p} < \infty,$$

then $f \in W_p^1(\mathbb{R}^n)$. To do this we use the following result:

Let ρ be a non-negative, radial function belonging to $L_1(\mathbb{R}^n)$, $g \in L_1(\mathbb{R}^n)$, $\phi \in C_0^2(\mathbb{R}^n)$ and $\mathbf{e} \in \mathbb{R}^n$, $|\mathbf{e}| = 1$. Then

$$\left| \int_{\mathbb{R}^n} g(x)\, dx \int_{(y-x)\cdot \mathbf{e} \ge 0} \frac{\phi(y)-\phi(x)}{|y-x|} \rho(y-x)\, dy \right|$$

$$\le \int_{\mathbb{R}^n} \int_{\mathbb{R}^n} \frac{|g(x)-g(y)|}{|x-y|} |\phi(y)|\, \rho(x-y)\, dx\, dy. \quad (3.2.12)$$

To prove this, let $\delta > 0$ and define

$$\rho^{(\delta)}(y) = \begin{cases} 0, & |y| < \delta, \\ \rho(y), & |y| > \delta. \end{cases}$$

It is enough to prove the result when ρ is replaced by $\rho^{(\delta)}$ and then allow $\delta \to 0$. Let

$$I := \int_{\mathbb{R}^n} g(x) \, dx \int_{(y-x)\cdot e \geq 0} \frac{\phi(y) - \phi(x)}{|y - x|} \rho^{(\delta)} (y - x) \, dy$$

$$= \int\int_{(y-x)\cdot e \geq 0} g(x)\phi(y) \frac{\rho^{(\delta)} (y - x)}{|y - x|} \, dx \, dy$$

$$- \int\int_{(y-x)\cdot e \geq 0} g(x)\phi(x) \frac{\rho^{(\delta)} (y - x)}{|y - x|} \, dx \, dy$$

$$:= I_1 - I_2,$$

the separation being justified since the integrands in I_1 and I_2 belong to $L_1 (\mathbb{R}^n \times \mathbb{R}^n)$. Interchange of x and y in I_2 and use of the radial property of $\rho^{(\delta)}$ shows that I_2 equals

$$\int\int_{(x-y)\cdot e \geq 0} g(y)\phi(y) \frac{\rho^{(\delta)} (x-y)}{|x-y|} \, dx \, dy = \int\int_{(y-x)\cdot e \geq 0} g(y)\phi(y) \frac{\rho^{(\delta)} (x-y)}{|x-y|} \, dx \, dy,$$

and so

$$I = \int\int_{(y-x)\cdot e \geq 0} \phi(y) \frac{g(x) - g(y)}{|y - x|} \rho^{(\delta)} (y - x) \, dx \, dy$$

$$\leq \int_{\mathbb{R}^n} \int_{\mathbb{R}^n} \frac{|g(x) - g(y)|}{|x - y|} |\phi(y)| \, \rho^{(\delta)} (x - y) \, dx \, dy;$$

thus (3.2.12) holds, as required. Note that the assumption that $g \in L_1 (\mathbb{R}^n)$ can be relaxed to $g \in L_{p,loc} (\mathbb{R}^n)$ as this weaker requirement is sufficient to ensure the existence of the various integrals because ϕ has compact support.

Finally, let $\phi \in C_0^\infty (\mathbb{R}^n)$, suppose that $e \in \mathbb{R}^n$, $|e| = 1$, and observe that arguments similar to those used when proving (3.2.11) show that for all $x \in \mathbb{R}^n$,

$$\int_{(y-x)\cdot e \geq 0} \frac{\phi(y) - \phi(x)}{|y - x|} \rho_\varepsilon(|y - x|) \, dy \to K \nabla \phi(x) \cdot e \qquad (3.2.13)$$

as $\varepsilon \to 0$, where

$$K = \frac{1}{2} \int_{\omega \in S^{n-1}} |\omega \cdot e| \, d\sigma := \frac{1}{2} K_{1,n}. \qquad (3.2.14)$$

We may apply (3.2.12) to $f \in L_p(\mathbb{R}^n)$ (see the comment in the proof of this result): together with Hölder's inequality and (3.2.5) this gives

$$J_\varepsilon := \left| \int_{\mathbb{R}^n} f(x)\, dx \int_{(y-x)\cdot e \geq 0} \frac{\phi(y) - \phi(x)}{|y-x|} \rho_\varepsilon(|y-x|)\, dy \right|$$

$$\leq \int_{\mathbb{R}^n} dx \int_{\mathrm{supp}\,\phi} \frac{|f(x) - f(y)|}{|x-y|} |\phi(y)|\, \rho_\varepsilon(|y-x|)\, dy$$

$$\leq \int_{\mathbb{R}^n} dx \int_{\mathbb{R}^n} \frac{|f(x) - f(y)|}{|x-y|} |\phi(y)|\, \rho_\varepsilon(|y-x|)\, dy$$

$$\leq \left(\int_{\mathbb{R}^n} \int_{\mathbb{R}^n} \frac{|f(x) - f(y)|^p}{|x-y|^p} \rho_\varepsilon(|y-x|)\, dx\, dy \right)^{1/p} \|\phi\|_{p'}.$$

Letting $\varepsilon \to 0$ we have

$$K \left| \int_{\mathbb{R}^n} f(x)\, (\nabla \phi(x) \cdot e)\, dx \right| \leq A_p \|\phi\|_{p'}.$$

The choice of e as the co-ordinate unit vector e_i $(i = 1, ..., n)$ shows that

$$\left| \int_{\mathbb{R}^n} f D_i \phi\, dx \right| \leq A_p \|\phi\|_{p'} / K,$$

so that $f \in W_p^1(\mathbb{R}^n)$. $\qquad\square$

Corollary 3.20 *Let $p \in (1, \infty)$. Then there is a constant K, depending only on p and n, such that for all $f \in L_p(\mathbb{R}^n)$ (and with the same understanding as in Theorem 3.19),*

$$\lim_{s \to 1-} (1-s) \int_{\mathbb{R}^n} \int_{\mathbb{R}^n} \frac{|f(x) - f(y)|^p}{|x-y|^{n+sp}}\, dx\, dy = \frac{K(p,n)}{p} \int_{\mathbb{R}^n} |\nabla f(x)|^p\, dx. \quad (3.2.15)$$

Proof We apply the last theorem with the particular choice of ρ_ε given by

$$\rho_\varepsilon(r) = \left\{ \begin{array}{ll} \frac{\varepsilon}{r^{n-\varepsilon}}, & 0 < r < 1, \\ 0, & r > 1 \end{array} \right\}.$$

This shows that

$$\lim_{\varepsilon \to 0} \varepsilon \int_{\mathbb{R}^n} \int_{|x-y|<1} \frac{|f(x) - f(y)|^p}{|x-y|^{n+p-\varepsilon}}\, dx\, dy = K(p,n) \int_{\mathbb{R}^n} |\nabla f(x)|^p\, dx.$$

Since

$$\int_{\mathbb{R}^n} \int_{|x-y|\geq 1} \frac{|f(x) - f(y)|^p}{|x-y|^{n+p-\varepsilon}}\, dx\, dy \leq C \|f\|_p^p$$

(see, for example, (3.2.1)), it follows that

$$\lim_{\varepsilon \to 0} \varepsilon \int_{\mathbb{R}^n} \int_{\mathbb{R}^n} \frac{|f(x) - f(y)|^p}{|x-y|^{n+p-\varepsilon}}\, dx\, dy = K(p,n) \int_{\mathbb{R}^n} |\nabla f(x)|^p\, dx,$$

which gives the corollary. $\qquad\square$

Remark 3.21 The proof of Theorem 3.19 given here is based on [23], which also deals with the case in which the functions are defined on a bounded domain Ω in $\mathbb{R}^n$ with smooth boundary, rather than on the whole of $\mathbb{R}^n$. Other approaches are given in [24] and [33]. Note that Corollary 3.20 immediately implies that if

$$\int_{\mathbb{R}^n} \int_{\mathbb{R}^n} \frac{|f(x) - f(y)|^p}{|x - y|^{n+sp}} \, dx \, dy = o\left(\frac{1}{1-s}\right) \text{ as } s \to 1-, \qquad (3.2.16)$$

then f is a constant function. The same holds when $\mathbb{R}^n$ is replaced by a smoothly bounded, connected open subset Ω of $\mathbb{R}^n$. In [33] Brezis shows that for any connected open subset Ω of $\mathbb{R}^n$, the condition

$$\int_{\Omega} \int_{\Omega} \frac{|f(x) - f(y)|^p}{|x - y|^{n+p}} \, dx \, dy < \infty \qquad (3.2.17)$$

is sufficient to ensure that f is constant. An elegant, simple proof of a result of this type is given in [158], pp. 214–215. A consequence is that the inequality

$$\left\| \frac{f(x) - f(y)}{|x - y|^{\frac{n}{p}+1}} \right\|_{L_p(\Omega \times \Omega)} \leq C(n)^{1/p} \|\nabla f\|_{L_p(\Omega)} \forall f \in C_0^\infty(\Omega)$$

does not hold. In [36] it is shown that there is a valid inequality

$$\left\| \frac{f(x) - f(y)}{|x - y|^{\frac{n}{p}+1}} \right\|_{L_{p,\infty}(\mathbb{R}^n \times \mathbb{R}^n)} \leq C(n)^{1/p} \|\nabla f\|_{L_p(\mathbb{R}^n)}, \qquad (3.2.18)$$

where $L_{p,\infty}$ is the Marcinkiewicz (or weak L_p) space. A natural question addressed in [35] is whether an improvement of (3.2.18) is possible in the Lorentz scale $L_{p,q}$, where $1 \leq p < \infty$ and $1 \leq q \leq \infty$. For $q \in (p, \infty)$ the answer is proved to be negative and there is the generalisation of (3.2.17)

$$\left\| \frac{f(x) - f(y)}{|x - y|^{\frac{n}{p}+1}} \right\|_{L_{p,q}(\mathbb{R}^n \times \mathbb{R}^n)} \leq C(n, p, q) \|\nabla f\|_{L_p(\mathbb{R}^n)} < \infty \Rightarrow f \text{ is a constant.}$$

$$(3.2.19)$$

The case when $s \to 0+$ was settled in [134], where it was shown that for all $f \in \cup_{0<s<1} W_p^s(\mathbb{R}^n)$,

$$\lim_{s \to 0+} s \int_{\mathbb{R}^n} \int_{\mathbb{R}^n} \frac{|f(x) - f(y)|^p}{|x - y|^{n+sp}} \, dx \, dy = C'(n, p) \int_{\mathbb{R}^n} |f(x)|^p \, dx.$$

3.3 An Approach via Interpolation Theory

Here we indicate how the limiting behaviour of the Gagliardo seminorm mentioned in Corollary 3.20 and Remark 3.21 may be explained as a consequence of interpolation theory. The approach we follow owes much to [107], [109] and [136]. To begin with, we recall some basic ideas.

A pair (A_0, A_1) of Banach spaces with norms $\|\cdot|A_0\|$, $\|\cdot|A_1\|$ is said to be *compatible* if there is a Hausdorff topological vector space A in which both A_0 and A_1 are continuously embedded. This implies that the sum $A_0 + A_1$ and the intersection $A_0 \cap A_1$ are well defined; endowed with the norms defined by

$$\|a|A_0 + A_1\| = \inf\{\|a_0|A_0\| + \|a_1|A_1\| : a = a_0 + a_1, a_i \in A_i \ (i = 0, 1)\}$$

and

$$\|a|A_0 \cap A_1\| = \max\{\|a|A_0\|, \|a|A_1\|\},$$

respectively, they are Banach spaces. Given such a compatible pair, $t \in (0, \infty)$ and $a \in A_0 + A_1$, the *K-functional* $K(t, a; A_0, A_1)$ is defined to be

$$\inf\{\|a_0|A_0\| + t\|a_1|A_1\| : a = a_0 + a_1, a_i \in A_i \ (i = 0, 1)\}. \tag{3.3.1}$$

Note that for all $t \in (0, \infty)$ and all $a \in A_0 + A_1$,

$$K(t, a; A_0, A_1) = tK(1/t, a; A_1, A_0). \tag{3.3.2}$$

Moreover (see, for example [21], Proposition 5.1.2), on $(0, \infty)$ the map $t \mapsto K(t, a; A_0, A_1)$ is increasing and $t \mapsto t^{-1}K(t, a; A_0, A_1)$ is decreasing. Thus

$$\lim_{t \to 0+} t^{-1}K(t, a; A_0, A_1) \text{ and } \lim_{t \to \infty} K(t, a; A_0, A_1)$$

exist (and may be infinite).

Let (A_0, A_1) be a compatible pair of Banach spaces; suppose that $s \in (0, 1)$ and $q \in [1, \infty)$. Then

$$(A_0, A_1)_{s,q} := \{a \in A_0 + A_1 : \|a|(A_0, A_1)_{s,q}\| < \infty\},$$

where

$$\|a|(A_0, A_1)_{s,q}\| := \left(\int_0^\infty \left(t^{-s}K(t, a; A_0, A_1)\right)^q \frac{dt}{t}\right)^{1/q}.$$

The space $(A_0, A_1)_{s,\infty}$ is defined analogously, with

$$\|a|(A_0, A_1)_{s,\infty}\| := \sup_{0 < t < \infty} t^{-s}K(t, a; A_0, A_1).$$

These are Banach spaces when equipped with the norms $\|\cdot|(A_0, A_1)_{s,q}\|$ and are said to be of *real interpolation type*. If no ambiguity is likely we shall simply write $K(t, a)$, $\|\cdot\|_{s,q}$, etc. We summarise some of the important properties of these interpolation spaces in the following theorem, giving indications of proofs for the convenience of the reader.

Theorem 3.22 *Let* (A_0, A_1), (B_0, B_1) *be compatible pairs of Banach spaces, let* $s \in (0, 1)$ *and suppose that* $q \in [1, \infty]$. *Then:*

(i) $(A_0, A_1)_{s,q} = (A_1, A_0)_{1-s,q}$.
(ii) *There is a constant* $c = c(s, q)$ *such that for all* $a \in (A_0, A_1)_{s,q}$ *and all* $t \in (0, \infty)$,

$$K(t, a) \le ct^s \|a\|_{s,q}.$$

(iii) *If* $q \le p \le \infty$, *then*

$$(A_0, A_1)_{s,1} \hookrightarrow (A_0, A_1)_{s,q} \hookrightarrow (A_0, A_1)_{s,p} \hookrightarrow (A_0, A_1)_{s,\infty}.$$

(iv) $A_0 = (A_0, A_0)_{s,q}$ *and for all* $a \in A_0$,

$$\|a_0|A_0\| = (s(1 - s)q)^{1/q} \|a|(A_0, A_0)_{s,q}\|.$$

(v)

$$A_0 \cap A_1 \hookrightarrow (A_0, A_1)_{s,q} \hookrightarrow A_0 + A_1.$$

(vi) *Let* $T: A_0 + A_1 \to B_0 + B_1$ *be linear and such that* $T|_{A_0} \in B(A_0, B_0)$ *and* $T|_{A_1} \in B(A_1, B_1)$. *Then*

$$\left\| T|B\left((A_0, A_1)_{s,q}, (B_0, B_1)_{s,q}\right) \right\| \le \|T|B(A_0, B_0)\|^{1-s} \|T|B(A_1, B_1)\|^s.$$

(vii) *There exists* $c = c(s, q)$ *such that for all* $a \in A_0 \cap A_1$,

$$\left\| a|(A_0, A_1)_{s,q} \right\| \le c \|a|A_0\|^{1-s} \|a|A_1\|^s.$$

Proof (i) This follows immediately from (3.3.2).

(ii) If $q < \infty$, then since $t \longmapsto K(t, a)$ is obviously monotonic increasing,

$$t^{-s} K(t, a) = (sq)^{1/q} K(t, a) \left(\int_t^\infty u^{-sq} \frac{du}{u} \right)^{1/q} \le (sq)^{1/q} \|a\|_{s,q}.$$

When $q = \infty$ the result is trivial.

(iii) It is enough to deal with the case in which $q < p < \infty$. By (ii), for all $a \in (A_0, A_1)_{s,q}$,

$$\|a\|_{s,p} \le \left(\int_0^\infty \left(t^{-s} K(t, a) \right)^q \frac{dt}{t} \right)^{1/p} \left(\sup_t t^{-s} K(t, a) \right)^{1-q/p}$$

$$\le c \|a\|_{s,q},$$

and (iii) follows.

(iv) Let $a \in A_0$. It is easy to see that

$$K(t, a; A_0, A_0) = \begin{cases} t \|a|A_0\|, & 0 \le t \le 1, \\ \|a|A_0\|, & 1 < t < \infty. \end{cases}$$

From this (iv) is immediate.

(v) This is a direct consequence of (ii) and the fact that

$$K(t, a) \leq \min \{1, t\} \, \|a|A_0 \cap A_1\|, \quad a \in A_0 \cap A_1.$$

(vi) If $T|_{A_0} \neq 0$,

$$K(t, Ta; B_0, B_1) \leq \inf_{a=a_0+a_1} \left(\|Ta_0\|_{B_0} + t \|Ta_1\|_{B_1} \right)$$

$$\leq \|T|B(A_0, B_0)\| \, K \left(\frac{\|T|B(A_1, B_1)\|}{\|T|B(A_0, B_0)\|} t, a; A_0, A_1 \right).$$

The transformation $\tau = t \, \|T|B(A_1, B_1)\| \, / \, \|T|B(A_0, B_0)\|$ now leads to the result. If $T|_{A_0} = 0$, replacement of $\|T|B(A_0, B_0)\|$ in the above argument by an arbitrarily small $\varepsilon > 0$ followed by passage of ε to 0 completes the argument.

(vii) See [170], p. 27. $\qquad\qquad\qquad\qquad\qquad\qquad\qquad\qquad\qquad\qquad\qquad \square$

As pointed out in Remark 3.13, when $s > 0, s \notin \mathbb{N}, p \in [1, \infty)$ and Ω is either the whole of $\mathbb{R}^n$ or a bounded open subset of $\mathbb{R}^n$ with smooth boundary, the space $W_p^s(\Omega)$ coincides with the Besov space $B_{p,p}^s(\Omega)$. The behaviour of the fractional Sobolev spaces under real interpolation can thus be deduced from that of Besov spaces, namely that if $p_0, p_1 \in [1, \infty), \theta \in (0, 1)$ and $s_0, s_1 \in (0, \infty)$, while $s = (1 - \theta)s_0 + \theta s_1, 1/p = (1 - \theta)/p_0 + \theta/p_1$, then (see [169], 3.3.6)

$$\left(B_{p_0,p_0}^{s_0}(\Omega), B_{p_1,p_1}^{s_1}(\Omega) \right)_{\theta,p} = B_{p,p}^s(\Omega).$$

A compatible pair (A_0, A_1) is said to be *normal* if

$$\lim_{t \to 0+} t^{-1} K(t, f; A_0, A_1) = \|f|A_1\| \text{ for all } f \in A_1, \qquad (3.3.3)$$

and

$$\lim_{t \to \infty} K(t, f; A_0, A_1) = \|f|A_0\| \text{ for all } f \in A_0. \qquad (3.3.4)$$

It is called *quasi-normal* if

$$\lim_{t \to 0+} t^{-1} K(t, \cdot; A_0, A_1) \text{ is equivalent to the norm } \|\cdot|A_1\| \text{ on } A_1,$$

and

$$\lim_{t \to \infty} K(t, \cdot; A_0, A_1) \text{ is equivalent to the norm } \|\cdot|A_0\| \text{ on } A_0.$$

These notions are key in what follows.

Theorem 3.23 *Let (A_0, A_1) be normal. Then:*

(i) If $1 \le q < \infty$ *and* $f \in A_0 \cap A_1$,

$$\lim_{s \to 1-} (qs(1-s))^{1/q} \|f\|_{s,q} = \|f|A_1\|$$

and

$$\lim_{s \to 0+} (qs(1-s))^{1/q} \|f\|_{s,q} = \|f|A_0\|.$$

(ii) If $1 \le q < \infty$ *and* $f \in A_0 \cap U_{s \in (0,1)} (A_0, A_1)_{s,q}$, *then*

$$\lim_{s \to 0+} (qs(1-s))^{1/q} \|f\|_{s,q} = \|f\|_{A_0}.$$

Proof   (i) First consider the case when $s \to 1-$. Given $\varepsilon > 0$, by (3.3.3) there exists $\delta > 0$ such that

$$\left| \left(\frac{K(t,f;A_0,A_1)}{t} \right)^q - \|f|A_1\|^q \right| < \varepsilon \text{ if } 0 < t < \delta. \qquad (3.3.5)$$

Note that

$$\left| qs(1-s) \|f\|_{s,q}^q - \|f|A_1\|^q \right| = \left| qs(1-s) \int_0^\infty \left(t^{-s} K(t,f) \right)^q \frac{dt}{t} - \|f|A_1\|^q \right|;$$

we write $\|f|A_1\|^q$ in the form

$$\|f|A_1\|^q q(1-s)\delta^{-(1-s)q} \int_0^\delta t^{(1-s)q} \frac{dt}{t}.$$

With $\int_0^\infty = \int_0^\delta + \int_\delta^\infty$ we thus obtain

$$\left| qs(1-s) \|f\|_{s,q}^q - \|f|A_1\|^q \right| \le I_1 + I_2 + I_3,$$

where

$$I_1 = qs(1-s) \int_0^\delta \left| t^{(1-s)q} \left(\left(\frac{K(t,f)}{t} \right)^q - \|f|A_1\|^q \right) \right| \frac{dt}{t},$$

$$I_2 = qs(1-s) \int_0^\delta \left| t^{(1-s)q} \left(\delta^{-(1-s)q} \frac{\|f|A_1\|^q}{s} - \|f|A_1\|^q \right) \right| \frac{dt}{t}$$

and

$$I_3 = qs(1-s) \int_\delta^\infty t^{(1-s)q} \left(\frac{K(t,f)}{t} \right)^q \frac{dt}{t}.$$

Use of (3.3.5) shows that

$$I_1 \le \varepsilon \delta^{(1-s)q} s, \qquad (3.3.6)$$

while plainly

$$I_2 = s \left| \left(s^{-1} - \delta^{-(1-s)q} \right) \delta^{(1-s)q} \right| \|f|A_1\|^q = \|f|A_1\|^q s \left| s^{-1} \delta^{(1-s)q} - 1 \right|. \qquad (3.3.7)$$

Since $K(t, f) \le \|f|A_0\|$ we see that

$$I_3 \le \delta^{-sq}(1 - s) \|f|A_0\|^q. \tag{3.3.8}$$

From these estimates it follows that if $1 - s$ is small enough, then

$$\left| qs(1 - s) \|f\|_{s,q}^q - \|f|A_1\|^q \right| < \varepsilon,$$

and the first part of (i) is established.

For the second part, we know from Theorem 3.22 (i) that $(A_1, A_0)_{1-s,q} = (A_0, A_1)_{s,q}$ and $\|f|(A_1, A_0)_{1-s,q}\| = \|f|(A_0, A_1)_{s,q}\|$, from which the rest of (i) follows.

It remains to prove (ii). Let $f \in A_0 \cap (A_0, A_1)_{s,q}$. Using the hypothesis of normality we see that, given $\varepsilon > 0$, there exists $\delta > 0$ such that

$$\left| K(t, f; A_0, A_1)^q - \|f|A_0\|^q \right| < \varepsilon \text{ when } t > \delta. \tag{3.3.9}$$

Moreover

$$\left| s(1 - s)q \|f\|_{s,q}^q - \|f|A_0\|^q \right|$$
$$= \left| s(1 - s)q \left\{ \int_0^\infty \left(t^{-s} K(t, f; A_0, A_1) \right)^q \frac{dt}{t} \right\} - \|f|A_0\|^q \right|$$
$$\le J_1 + J_2 + J_3,$$

where

$$J_1 = s(1 - s)q \int_0^\delta \left(t^{-s} K(t, f; A_0, A_1) \right)^q \frac{dt}{t},$$

$$J_2 = s(1 - s)q \int_\delta^\infty t^{-sq} \left| K(t, f; A_0, A_1)^q - \|f|A_0\|^q \right| \frac{dt}{t}$$

and

$$J_3 = \|f|A_0\|^q \left((1 - s) \delta^{sq} - 1 \right).$$

Suppose that $s < s_0$. Then

$$J_1 = s(1 - s)q\delta^{-sq} \int_0^\delta \left((t/\delta)^{-s} K(t, f; A_0, A_1) \right)^q \frac{dt}{t}$$
$$\le s(1 - s)q\delta^{-sq} \int_0^\delta \left((t/\delta)^{-s_0} K(t, f; A_0, A_1) \right)^q \frac{dt}{t}$$
$$\le s(1 - s)q\delta^{q(s_0-s)} \int_0^\delta \left(t^{-s_0} K(t, f; A_0, A_1) \right)^q \frac{dt}{t}$$
$$\le s(1 - s)q\delta^{q(s_0-s)} \|f\|_{s_0,q}^q,$$

and so $\lim_{s \to 0} J_1 = 0$. Using (3.3.9) we see that $J_2 < \varepsilon$ for small enough s; that $\lim_{s \to 0} J_3 = 0$ is clear. $\qquad\square$

Remark 3.24 If the requirement of normality is weakened to quasi-normality, minor changes to the proof show that a modified form of the theorem still holds. More precisely, (i) and (ii) hold with $\|f|A_0\|$ and $\|f|A_1\|$ replaced by $\lim_{t\to\infty} K(t,f;A_0,A_1)$ and $\lim_{t\to 0+} t^{-1}K(t,f;A_0,A_1)$, respectively.

We can now apply this abstract result to fractional Sobolev spaces. Let $p \in [1,\infty)$ and denote by $W^1_{p,0}(\mathbb{R}^n)$ the completion of $C^\infty_0(\mathbb{R}^n)$ with respect to the norm given by

$$\|f|W^1_{p,0}(\mathbb{R}^n)\| := \||\nabla f|\|_{p,\mathbb{R}^n}.$$

With the understanding that functions in $W^1_{p,0}(\mathbb{R}^n)$ that differ by a constant are identified, the pair $\left(L_p(\mathbb{R}^n), W^1_{p,0}(\mathbb{R}^n)\right)$ of Banach spaces is compatible. The K-functional for this pair satisfies

$$K\left(t,f;L_p(\mathbb{R}^n),W^1_{p,0}(\mathbb{R}^n)\right) \approx w_p(f,t), \quad t > 0, \tag{3.3.10}$$

where

$$w_p(f,t) := \sup_{|h|\le t} \|\Delta_h f\|_{p,\mathbb{R}^n}, \quad \Delta_h f(x) := f(x+h) - f(x). \tag{3.3.11}$$

For this we refer to [21], p. 341; the constants of equivalence are independent of f and t. Denoting by $X_{s,p}$ the interpolation space $\left(L_p(\mathbb{R}^n), W^1_{p,0}(\mathbb{R}^n)\right)_{s,p}$, it follows that

$$\|f|X_{s,p}\| \approx \left(\int_0^\infty \left(t^{-s}w_p(f,t)\right)^p \frac{dt}{t}\right)^{1/p}. \tag{3.3.12}$$

Lemma 3.25 *Let $s \in (0,1)$, $p \in [1,\infty)$ and denote by $W^s_{p,0}(\mathbb{R}^n)$ the completion of $C^\infty_0(\mathbb{R}^n)$ with respect to the norm given by*

$$\|f|W^s_{p,0}(\mathbb{R}^n)\| := \left(\int_{\mathbb{R}^n}\int_{\mathbb{R}^n} \frac{|f(x)-f(y)|^p}{|x-y|^{n+sp}}\,dx\,dy\right)^{1/p} = [f]_{s,p,\mathbb{R}^n}.$$

Then

$$X_{s,p} = \left(L_p(\mathbb{R}^n), W^1_{p,0}(\mathbb{R}^n)\right)_{s,p} = W^s_{p,0}(\mathbb{R}^n) \tag{3.3.13}$$

and

$$\|f|W^s_{p,0}(\mathbb{R}^n)\| \approx (n+sp)^{1/p}\|f|X_{s,p}\|. \tag{3.3.14}$$

Proof Of course $\left\|\cdot|W^s_{p,0}(\mathbb{R}^n)\right\|$ is just the Gagliardo seminorm, which is a norm in this context. Also $W^s_{p,0}(\mathbb{R}^n)$ coincides with the space $\overset{0}{\mathcal{D}}^s_p(\mathbb{R}^n)$ mentioned in Proposition 3.7; the present notation is used here as it is more suggestive in this context. We use the fact that

$$w_p(f, t) \approx \left(t^{-n} \int_{|h| \leq t} \| \Delta_h f \|_{p, \mathbb{R}^n}^p \, dh \right)^{1/p}. \tag{3.3.15}$$

Assuming this for the moment, Fubini's theorem shows that

$$\int_0^\infty \left(t^{-s} w_p(f, t) \right)^p \frac{dt}{t} \approx (n + sp)^{-1} \int_{\mathbb{R}^n} \int_{\mathbb{R}^n} \frac{|f(x) - f(y)|^p}{|x - y|^{n+sp}} \, dx \, dy,$$

from which the result follows.

To establish (3.3.15), observe that the estimate of $w_p(f, t)$ from below is obvious. For the upper bound write, for $p \in (1, \infty)$ and $q \in (0, \infty]$,

$$\tau_q(f, t) = \left\{ t^{-n} \int_{|h| \leq t} \| \Delta_h f \|_{p, \mathbb{R}^n}^q \, dh \right\}^{1/q} \text{ if } 0 < q < \infty, \quad \tau_\infty(f, t) = w_p(f, t).$$

Plainly

$$\tau_q(f, t) \leq |B(0, 1)|^{1/q - 1/r} \tau_r(f, t) \text{ if } q < r < \infty.$$

It is thus enough to show that there is a constant $c = c(q)$ such that

$$\tau_\infty(f, t) \leq c(q) \tau_q(f, t), 0 < q < 1.$$

To do this, let $|h|, |\xi| \leq t$ and note that

$$\| \Delta_h f \|_p \leq \| \Delta_{\xi - h} f \|_p + \| \Delta_\xi f \|_p \leq 2 \| \Delta_{(\xi - h)/2} f \|_p + \| \Delta_\xi f \|_p.$$

Since $|(\xi - h)/2| \leq t$,

$$\| \Delta_h f \|_p^q \leq 2^q \| \Delta_{(\xi - h)/2} f \|_p^q + \| \Delta_\xi f \|_p^q$$

and

$$\int_{|\xi| \leq t} \| \Delta_h f \|_p^q \, d\xi \leq (2^q + 1) \int_{|\xi| \leq t} \| \Delta_\xi f \|_p^q \, d\xi.$$

Hence there is a constant $C = C(q)$ such that

$$\| \Delta_h f \|_p \leq C \tau_q(f, t), 0 < q < 1, |h| \leq t,$$

and the proof is complete. $\qquad \square$

Lemma 3.26 *The pair* $\left(L_p(\mathbb{R}^n), W_{p,0}^1(\mathbb{R}^n) \right) (p \in [1, \infty))$ *is quasi-normal.*

Proof Let $f \in W_{p,0}^1(\mathbb{R}^n)$. Then for each $h \in \mathbb{R}^n$ and almost every $x \in \mathbb{R}^n$,

$$f(x + h) - f(x) = \int_0^1 \nabla f(x + th) \cdot h \, dt$$

(see Proposition 2.6). Then

$$\left(\int_{\mathbb{R}^n} |f(x+h) - f(x)|^p \, dx \right)^{1/p} \leq |h| \, \|\nabla f\|_p \, ,$$

so that (see (3.3.11))

$$w_p(f, \delta) \leq \delta \, \|\nabla f\|_p \, . \tag{3.3.16}$$

Given $\varepsilon > 0$, there exists $f_\varepsilon \in C_0^\infty(\mathbb{R}^n)$ such that $\|\nabla (f - f_\varepsilon)\|_p < \varepsilon$. Moreover,

$$\mu_\varepsilon(\delta) := \sup_{|h|=\delta} \|f_\varepsilon(\cdot + h) - f_\varepsilon(\cdot) - \nabla f_\varepsilon(\cdot) \cdot h\|_p / \delta \to 0 \text{ as } \delta \to 0,$$

and hence there exists $\delta_\varepsilon > 0$ such that $\mu_\varepsilon(\delta) < \varepsilon$ if $0 < \delta < \delta_\varepsilon$. By (3.3.16),

$$w_p(f_\varepsilon, \delta) \leq w_p(f, \delta) + \varepsilon\delta.$$

Thus for all $\delta \in (0, \delta_\varepsilon)$,

$$\|\nabla f\|_p \leq \|\nabla f_\varepsilon\|_p + \varepsilon \leq \mu_\varepsilon(\delta) + \delta^{-1} w_p(f_\varepsilon, \delta) + \varepsilon$$
$$\leq \delta^{-1} w_p(f, \delta) + 3\varepsilon.$$

Hence

$$\lim_{\delta \to 0+} \frac{w_p(f, \delta)}{\delta} = \|\nabla f\|_p \, . \tag{3.3.17}$$

Moreover, $t \longmapsto w_p(f, t)$ is monotonic increasing and $w_p(f, t) \leq 2 \|f\|_p$; hence $\lim_{t \to \infty} w_p(f, t)$ exists and is bounded above by $2 \|f\|_p$. To obtain a lower bound, let $f \in C_0^\infty(\mathbb{R}^n)$, and suppose $\operatorname{supp} f \subset \{x \in \mathbb{R}^n : |x| < M\}$, so that $f(x+h) = 0$ if $|h| \geq 2M$ and $|x| < M$. Hence

$$w_p(f, t) \geq \|f\|_p \text{ if } t \geq 2M,$$

and so

$$\|f\|_p \leq \lim_{t \to \infty} w_p(f, t) \leq 2 \|f\|_p \, .$$

If $f \in L_p(\mathbb{R}^n)$, then let $(f_k)_{k \in \mathbb{N}} \subset C_0^\infty(\mathbb{R}^n)$ be such that $\lim_{k \to \infty} \|f - f_k\|_p = 0$ and observe that

$$\|f_k\|_p \leq \lim_{t \to \infty} w_p(f_k, t) \leq \lim_{t \to \infty} w_p(f_k - f, t) + \lim_{t \to \infty} w_p(f, t)$$
$$\leq 2 \|f - f_k\|_p + \lim_{t \to \infty} w_p(f, t),$$

which shows that

$$\|f\|_p \leq \lim_{t \to \infty} w_p(f, t),$$

and completes the proof. $\qquad\qquad\square$

Application of this result to the pair $\left(L_p\left(\mathbb{R}^n\right), W_{p,0}^1\left(\mathbb{R}^n\right)\right)$ is thus possible. Let $f \in W_{p,0}^1\left(\mathbb{R}^n\right)$. Then

$$\lim_{s\to 1-} s^{-1/p} p^{-1/p} (1-s)^{1/p} \left\{ \int_{\mathbb{R}^n} \int_{\mathbb{R}^n} \frac{|f(x) - f(y)|^p}{|x-y|^{n+sp}} \, dx \, dy \right\}^{1/p}$$

coincides with

$$\lim_{s\to 1-} s^{-1/p} p^{-1/p} (1-s)^{1/p} \left\| f | W_{p,0}^s\left(\mathbb{R}^n\right) \right\|$$

$$\approx \lim_{s\to 1-} (n+sp)^{1/p} s^{-1/p} p^{-1/p} \left\| f | \left(L_p\left(\mathbb{R}^n\right), W_{p,0}^1\left(\mathbb{R}^n\right)\right)_{s,p} \right\|$$

$$\approx (n+p)^{1/p} p^{-1/p} \left\| f | W_{p,0}^1\left(\mathbb{R}^n\right) \right\|$$

$$= (n+p)^{1/p} p^{-1/p} \left\| \nabla f \right\|_p,$$

in line with the Bourgain, Brezis, Mironescu result of [24] insofar as the rate of blow up is concerned, but without the exact constant that they obtain.

In the direction of the Maz'ya–Shaposhnikova theorem of [134] we observe that if $f \in \cup_{s\in(0,1)} W_{p,0}^s\left(\mathbb{R}^n\right)$, then

$$\lim_{s\to 0+} s \left\{ \int_{\mathbb{R}^n} \int_{\mathbb{R}^n} \frac{|f(x) - f(y)|^p}{|x-y|^{n+sp}} \, dx \, dy \right\} = \lim_{s\to 0+} s \left\| f | W_{p,0}^s\left(\mathbb{R}^n\right) \right\|^p$$

$$\approx p^{-1} n \lim_{s\to 0+} \left\| f | \left(L_p\left(\mathbb{R}^n\right), W_{p,0}^1\left(\mathbb{R}^n\right)\right)_{s,p} \right\|^p$$

$$= p^{-1} n \left\| f \right\|_p^p.$$

The paper [31] contains much interesting information about the interpolation approach to the fractional spaces.

3.4 Connections with the Laplacian

Definition 3.27 Given any $s \in (0, 1)$, the corresponding fractional power of the Laplacian is the map $(-\Delta)^s \colon \mathcal{S} \to L_2\left(\mathbb{R}^n\right)$ given by

$$(-\Delta)^s u = F^{-1}\left(|\xi|^{2s} F(u)\right). \tag{3.4.1}$$

Since the Fourier transform F maps the Schwartz space $\mathcal{S}$ onto itself, it is easy to see, using the dominated convergence theorem, that

$$\lim_{s\to 0+} (-\Delta)^s u(x) = u(x) \text{ and } \lim_{s\to 1-} (-\Delta)^s u(x) = -\Delta u(x) \ (u \in \mathcal{S}, x \in \mathbb{R}^n).$$

Note that since $\mathcal{S}$ is dense in every space $H^t\left(\mathbb{R}^n\right)$ with $t > 0$, we may and shall suppose that this definition holds for all $u \in H^t\left(\mathbb{R}^n\right)$. We first establish a connection between this fractional power and the corresponding Gagliardo seminorm $[u|H^s(\mathbb{R}^n)] = [u|W_2^s(\mathbb{R}^n)] := [u]_{s,2,\mathbb{R}^n}$ (see Section 3.1).

Proposition 3.28 *Let $s \in (0, 1)$. Then for all $u \in H^s(\mathbb{R}^n)$,*

$$[u|H^s(\mathbb{R}^n)]^2 = 2C(n,s)^{-1} \left\| (-\Delta)^{s/2} u \right\|_2^2, \qquad (3.4.2)$$

where

$$C(n,s) = \left(\int_{\mathbb{R}^n} \frac{1-\cos\zeta_1}{|\zeta|^{n+2s}} d\zeta \right)^{-1} = 2^{2s} \pi^{-n/2} \frac{\Gamma\left(\frac{n}{2}+s\right)}{|\Gamma(-s)|}. \qquad (3.4.3)$$

Proof Using a change of variable and the Plancherel formula we see that

$$[u|H^s(\mathbb{R}^n)]^2 = \int_{\mathbb{R}^n} \left(\int_{\mathbb{R}^n} \frac{|u(x)-u(y)|^2}{|x-y|^{n+2s}} dx \right) dy \qquad (3.4.4)$$

$$= \int_{\mathbb{R}^n} \int_{\mathbb{R}^n} \left(\frac{|u(z+y)-u(y)|^2}{|z|^{n+2s}} dy \right) dz$$

$$= \int_{\mathbb{R}^n} \left\| \frac{u(z+\cdot)-u(\cdot)}{|z|^{n/2+s}} \right\|_{2,\mathbb{R}^n}^2 dz$$

$$= \int_{\mathbb{R}^n} \left\| F\left(\frac{u(z+\cdot)-u(\cdot)}{\left|z^{n/2+s}\right|} \right) \right\|_{2,\mathbb{R}^n}^2 dz$$

$$= \int_{\mathbb{R}^n} \int_{\mathbb{R}^n} \frac{\left|e^{i\xi\cdot z}-1\right|^2}{|z|^{n+2s}} |Fu(\xi)|^2 \, dz\,d\xi$$

$$= 2 \int_{\mathbb{R}^n} \int_{\mathbb{R}^n} \frac{(1-\cos(\xi\cdot z))}{|z|^{n+2s}} |Fu(\xi)|^2 \, dz\,d\xi. \qquad (3.4.5)$$

We claim that the function $G\colon \mathbb{R}^n \to \mathbb{R}$ defined by

$$G(\xi) = \int_{\mathbb{R}^n} \frac{(1-\cos(\xi\cdot z))}{|z|^{n+2s}} dz$$

is given by

$$G(\xi) = C(n,s)^{-1} |\xi|^{2s}.$$

To see this, observe that G is rotationally invariant: $G(\xi) = G(|\xi|e_1)$, where $e_1(1, 0, ..., 0) \in \mathbb{R}^n$. This is obvious when $n = 1$. When $n \geq 2$, let R be the rotation for which $R(|\xi|e_1) = \xi$, denote its transpose by R^T, put $y = R^T z$ and note that

$$G(\xi) = \int_{\mathbb{R}^n} \frac{1 - \cos\left((R\left(|\xi| e_1\right)) \cdot z\right)}{|z|^{n+2s}} \, dz$$

$$= \int_{\mathbb{R}^n} \frac{1 - \cos\left((|\xi| e_1) \cdot (R^T z)\right)}{|z|^{n+2s}} \, dz$$

$$= \int_{\mathbb{R}^n} \frac{1 - \cos(|\xi| e_1 \cdot y)}{|y|^{n+2s}} \, dy = G\left(|\xi| e_1\right),$$

as claimed. Hence, with $\zeta = |\xi| y$, we have

$$G(\xi) = G\left(|\xi| e_1\right) = \int_{\mathbb{R}^n} \frac{1 - \cos\left(|\xi| y_1\right)}{|y|^{n+2s}} \, dy$$

$$= \frac{1}{|\xi|^n} \int_{\mathbb{R}^n} \frac{1 - \cos \zeta_1}{|\zeta/|\xi||^{n+2s}} d\zeta = C(n,s)^{-1} |\xi|^{2s},$$

as desired. Together with (3.4.5) this shows that

$$\left[u|H^s\left(\mathbb{R}^n\right)\right]^2 = 2C(n,s)^{-1} \||\xi|^s Fu\|_{2,\mathbb{R}^n}^2 = 2C(n,s)^{-1} \left\|(-\Delta)^{s/2} u\right\|_2^2.$$

The formula for $C(n,s)$ given in (3.4.3) follows from the fact that

$$C(n,s)^{-1} = \int_0^\infty \left(|S^{n-1}| - (2\pi)^{n/2} r^{-(n-2)/2} J_{(n-2)/2}(r)\right) r^{-2s-1} dr,$$

where $J_{(n-2)/2}$ is the Bessel function of the first kind of order $(n-2)/2$, and the identity

$$\int_0^\infty r^{-z} \left(J_{(n-2)/2}(r) - 2^{-(n-2)/2}\Gamma(n/2^{-1} r^{(n-2)/2}\right) dr = 2^{-z} \frac{\Gamma\left((n-2z)/4\right)}{\Gamma\left((n+2z)/4\right)}$$

for $n/2 < z < (n+4)/2$ given in (2.20) of [172]; see also [78], proof of Lemma 1. $\qquad\square$

Remark 3.29 From (3.4.3) it is plain that the constant $C(n,s)$ has the following behaviour:

$$\lim_{s\to 0+} s^{-1} C(n,s) = 2/\omega_n, \quad \lim_{s\to 1-} (1-s)^{-1} C(n,s) = 4n/\omega_n.$$

Proposition 3.30 *Let $s \in (0,1)$. Then for all $u \in S$,*

$$(-\Delta)^s u(x) = -\frac{1}{2}C(n,s) \int_{\mathbb{R}^n} \frac{u(x+y) + u(x-y) - 2u(x)}{|y|^{n+2s}} \, dy \ (x \in \mathbb{R}^n).$$

$$(3.4.6)$$

Proof Denote the right-hand side of (3.4.6) by $\mathcal{L}u(x)$. Use of Taylor's theorem shows that for each fixed $x \in \mathbb{R}^n$ and $y \in \mathbb{R}^n \setminus \{0\}$,

$$\frac{|u(x+y)+u(x-y)-2u(x)|}{|y|^{n+2s}} \leq C\,|y|^{2-n-2s} \sup\{|D^\alpha u(z)| : |\alpha| = 2, \ z \in B(x,1)\},$$

so that as $u \in \mathcal{S}$, $\mathcal{L}u(x)$ exists. Moreover, the condition $u \in \mathcal{S}$ also implies that for each fixed $y \in \mathbb{R}^n \backslash \{0\}$ the map $x \longmapsto |u(x+y) + u(x-y) - 2u(x)| \, |y|^{-n-2s}$ belongs to $L_1(\mathbb{R}^n)$. Since

$$-\frac{1}{2} C(n, s) \int_{\mathbb{R}^n} \frac{F\left(u(\cdot + y) + u(\cdot - y) - 2u(\cdot)\right)(\xi)}{|y|^{n+2s}} \, dy$$

$$= -\frac{1}{2} C(n, s) \int_{\mathbb{R}^n} \frac{e^{i\xi \cdot y} + e^{-i\xi \cdot y} - 2}{|y|^{n+2s}} (Fu)(\xi) \, dy$$

$$= C(n, s)(Fu)(\xi) \int_{\mathbb{R}^n} \frac{1 - \cos(\xi \cdot y)}{|y|^{n+2s}} \, dy = |\xi|^{2s} (Fu)(\xi),$$

the final step following from (3.4.3), we conclude from the Fubini–Tonelli theorem that $\mathcal{L}u(x) = F^{-1}\left(|\xi|^{2s}(Fu)(\xi)\right)(x) = (-\Delta)^s(x)$, as required. $\qquad\square$

Using this result we can now give yet another form of the fractional Laplacian. Let $u \in \mathcal{S}$ and note that

$$\lim_{\varepsilon \to 0^+} \int_{\mathbb{R}^n \backslash B(x,\varepsilon)} \frac{u(y) - u(x)}{|x-y|^{n+2s}} \, dy := pv \int_{\mathbb{R}^n} \frac{u(y) - u(x)}{|x-y|^{n+2s}} \, dy$$

$$= pv \int_{\mathbb{R}^n} \frac{u(x+z) - u(x)}{|z|^{n+2s}} \, dz = pv \int_{\mathbb{R}^n} \frac{u(x-z) - u(x)}{|z|^{n+2s}} \, dz,$$

so that

$$2pv \int_{\mathbb{R}^n} \frac{u(x+z) - u(x)}{|z|^{n+2s}} \, dz = pv \int_{\mathbb{R}^n} \frac{u(x+z) - u(x)}{|z|^{n+2s}} dz + pv \int_{\mathbb{R}^n} \frac{u(x-z) - u(x)}{|z|^{n+2s}} dz$$

$$= pv \int_{\mathbb{R}^n} \frac{u(x+z) + u(x-z) - 2u(x)}{|z|^{n+2s}} \, dz.$$

As observed above,

$$\left| \frac{u(x+z) + u(x-z) - 2u(x)}{|z|^{n+2s}} \right| \le |z|^{2-n-2s} \sup_{|\alpha|=2, y \in \mathbb{R}^n} |D^\alpha u(y)|$$

and since the right-hand side is integrable near 0, the pv can be removed and we have

$$pv \int_{\mathbb{R}^n} \frac{u(y) - u(x)}{|x-y|^{n+2s}} \, dy = \frac{1}{2} \int_{\mathbb{R}^n} \frac{u(x+z) + u(x-z) - 2u(x)}{|z|^{n+2s}} \, dz$$

$$= -C(n, s)^{-1} \mathcal{L}u(x),$$

giving

$$(-\Delta)^s(x) = C(n, s) pv \int_{\mathbb{R}^n} \frac{u(y) - u(x)}{|x-y|^{n+2s}} \, dy. \tag{3.4.7}$$

Companion to the developments just outlined involving fractional powers of the Laplacian there is the rapidly evolving theory of the fractional p-Laplacian.

To introduce this, suppose that Ω is a bounded open subset of $\mathbb{R}^n$ with smooth boundary and let $p \in (1, \infty)$. The p-Laplacian Δ_p can be defined by its action on smooth enough functions u:

$$\Delta_p u := \sum_{j=1}^{n} D_j \left(|\nabla u|^{p-2} D_j u \right).$$

It arises naturally on seeking to minimise the Rayleigh quotient

$$R(u) := \int_{\Omega} |\nabla u|^p \, dx / \int_{\Omega} |u|^p \, dx$$

among all functions $u \in C_0^\infty(\Omega) \backslash \{0\}$. We refer to [65], Chapter 9 for some details of the basic theory.

To obtain a fractional version of this suppose $s \in (0, 1)$ and let

$$X(\Omega) := \left\{ u \in W_p^s(\mathbb{R}^n) : u = 0 \text{ a.e. in } \mathbb{R}^n \backslash \Omega \right\};$$

this is endowed with the norm $[\cdot]_{s,p,\mathbb{R}^n}$ and then becomes a uniformly convex, uniformly smooth Banach space that coincides with the space $\overset{0}{\mathcal{D}_p^s}(\Omega)$ defined earlier. The nonlinear map $A: X(\Omega) \to X(\Omega)^*$ defined by

$$\langle Au, v \rangle = \int_{\mathbb{R}^n} \int_{\mathbb{R}^n} \frac{|u(x) - u(y)|^{p-2} \, (u(x) - u(y)) \, (v(x) - v(y))}{|x - y|^{n+sp}} \, dx \, dy$$

for all $u, v \in X(\Omega)$ is a duality map (see Section 1.2):

$$\langle Au, u \rangle = \|u|X(\Omega)\|^p \, , \, |\langle Au, v \rangle| \le \|u|X(\Omega)\|^{p-1} \, \|v|X(\Omega)\| \, .$$

This map A is called the s fractional p-Laplacian and is denoted by $(-\Delta)_p^s$; it is the gradient of the functional J defined by

$$J(u) = \frac{1}{p} \int_{\mathbb{R}^n} \int_{\mathbb{R}^n} \frac{|u(x) - u(y)|^p}{|x - y|^{n+sp}} \, dx \, dy = \frac{1}{p} [u]_{s,p,\mathbb{R}^n}^p \, . \tag{3.4.8}$$

To see how this operator is connected with a boundary-value problem, let

$$S = \{ u \in X(\Omega) : I(u) = 1 \}, \text{ where } I(u) := \|u\|_{p,\Omega}^p \, .$$

Then $I, J \in C^1(X(\Omega))$ and $\langle J'(u), v \rangle = \langle Au, v \rangle$ for all $u, v \in X(\Omega)$. Let $\widetilde{J}$ be the restriction of J to S. We claim that $\lambda > 0$ is a critical value of $\widetilde{J}$ if and only if it is an eigenvalue of the (weak) problem

$$(-\Delta)_p^s u = \lambda \, |u|^{p-2} u \text{ in } \Omega,$$

$$u = 0 \text{ in } \mathbb{R}^n \backslash \Omega.$$

Here by a weak solution of this problem is meant a function $u \in X(\Omega)$ such that for all $v \in X(\Omega)$,

$$\int_{\mathbb{R}^n} \int_{\mathbb{R}^n} \frac{|u(x) - u(y)|^{p-2} (u(x) - u(y)) (v(x) - v(y))}{|x - y|^{n+sp}} \, dx \, dy = \lambda \int_{\Omega} |u|^{p-2} \, uv \, dx.$$

For suppose that $u \in S$ and $\mu \in \mathbb{R}$ are such that $J(u) = \lambda$ and $J'(u) - \mu I'(u) = 0$ in $X(\Omega)^*$. Then for all $v \in X(\Omega)$,

$$\int_{\mathbb{R}^n} \int_{\mathbb{R}^n} \frac{|u(x) - u(y)|^{p-2} (u(x) - u(y)) (v(x) - v(y))}{|x - y|^{n+sp}} \, dx \, dy = \mu \int_{\Omega} |u|^{p-2} \, uv \, dx;$$

taking $v = u$, we see that $\lambda = \mu$ and $u \neq 0$ is a weak solution of the problem. Conversely, if λ is an eigenvalue of the weak problem, then there is a corresponding weak eigenfunction $u \in X(\Omega)$ with $I(u) = 1$. By Proposition 3.54 of [144], $u \in S$ is a critical point of $\widetilde{J}$ at level λ.

We next show that A has a property (introduced by Browder [37]) that will prove to be useful in our discussion of the spectrum. Given a Banach space Y, a map $T: Y \to Y^*$ is said to be *of type* $(S)_+$ if, whenever (u_k) is a sequence in Y such that $u_k \rightharpoonup u \in Y$ and

$$\limsup_{k \to \infty} \langle Tu_k - Tu, u_k - u \rangle \leq 0,$$

then $u_k \to u$ in Y. Following [151] we show that the map A discussed above has this property.

Lemma 3.31 *The map* $A: X(\Omega) \to X(\Omega)^*$ *is of type* $(S)_+$.

Proof Suppose that $u_k \rightharpoonup u$ in $X(\Omega)$ and

$$\limsup_{k \to \infty} \langle Au_k - Au, u_k - u \rangle \leq 0.$$

Since

$$\langle Au_k - Au, u_k - u \rangle \geq \|u_k|X(\Omega)\|^{p-1} (\|u_k|X(\Omega)\| - \|u|X(\Omega)\|) - \langle Au, u_k - u \rangle,$$

we see that

$$\lim_{k \to \infty} \langle Au_k - Au, u_k - u \rangle = 0.$$

Now we use Lemma 1.11. If $p \geq 2$, this shows that

$$\|u_k - u|X(\Omega)\|^p \leq C_p \langle Au_k - Au, u_k - u \rangle \to 0 \text{ as } k \to \infty,$$

while when $1 < p < 2$, $\|u_k - u|X(\Omega)\|^p$ is bounded above by

$$C_p^{p/2} \, |\langle Au_k - Au, u_k - u\rangle|^{p/2} \, (\|u_k|X(\Omega)\|^p + \|u|X(\Omega)\|^p)^{(2-p)/2}$$

$$\leq C_p^{p/2} \, |\langle Au_k - Au, u_k - u\rangle|^{p/2} \left(\|u_k|X(\Omega)\|^{p(2-p)/2} + \|u|X(\Omega)\|^{p(2-p)/2}\right)$$

$$\leq C \, |\langle Au_k - Au, u_k - u\rangle|^{p/2} \to 0 \text{ as } k \to \infty.$$

The lemma follows. $\qquad\qquad\qquad\qquad\qquad\qquad\qquad\qquad\qquad\qquad\qquad\quad\square$

For much information about the s fractional p-Laplacian and associated boundary-value problems we refer to [28], [29], [102] and [127].

4

Eigenvalues of the Fractional p-Laplacian

4.1 Fundamentals

Throughout we suppose that Ω is a bounded open subset of $\mathbb{R}^n$, let $p \in (1, \infty)$ and assume that $s \in (0, 1)$; the spaces involved are assumed to be real. Consider the problem of minimising the fractional Rayleigh quotient

$$R(f, p, s, \Omega) := \frac{\int_{\mathbb{R}^n} \int_{\mathbb{R}^n} \frac{|f(x) - f(y)|^p}{|x - y|^{n+sp}} \, dx \, dy}{\int_{\mathbb{R}^n} |f(x)|^p \, dx} = \frac{[f]^p_{s,p,\mathbb{R}^n}}{\|f\|^p_{p,\mathbb{R}^n}} \tag{4.1.1}$$

among all $f \in C_0^\infty(\Omega) \backslash \{0\}$. We write

$$\lambda_1 = \lambda_1(p, s, \Omega) = \inf \left\{ R(f, p, s, \Omega) : f \in C_0^\infty(\Omega) \backslash \{0\} \right\}.$$

Since Ω supports the (s, p)-Friedrichs inequality (see Proposition 3.17), it is clear that $\lambda_1 > 0$; we refer to λ_1 as the *first eigenvalue* of the fractional p-Laplacian $(-\Delta)_p^s$. The attainment of the infimum is discussed in the next theorem, which reinforces the treatment of Section 3.4. We recall that the space $X := \overset{0}{\mathcal{D}}_p^s(\Omega)$ that appears there and below is the completion of $C_0^\infty(\Omega)$ with respect to the norm $[\cdot]_{s,p,\mathbb{R}^n}$: see Proposition 3.7 for information about this space. In particular, note that X and X^* are uniformly convex and that X coincides with $\overset{0}{X}_p^s(\Omega)$ since Ω is bounded.

Theorem 4.1 *The infimum of R is attained at a non-negative function $f \in X \backslash \{0\}$, with $f = 0$ in $\mathbb{R}^n \backslash \Omega$. This minimising function f satisfies the Euler–Lagrange equation*

$$\int_{\mathbb{R}^n} \int_{\mathbb{R}^n} \frac{|f(x) - f(y)|^{p-2} (f(x) - f(y)) (\phi(x) - \phi(y))}{|x - y|^{n+sp}} \, dx \, dy = \lambda_1 \int_{\mathbb{R}^n} |f|^{p-2} f\phi \, dx \tag{4.1.2}$$

for all $\phi \in C_0^\infty(\Omega)$.

Proof   When minimising the Rayleigh quotient it is enough to consider non-negative functions since for every $u \in L_p(\Omega)$,

$$||u(x)| - |u(y)||^p \leq |u(x) - u(y)|^p \text{ and } \||u|\|_p = \|u\|_p \,.$$

Let $\{f_j\}_{j \in \mathbb{N}}$ be a minimising sequence, with $f_j \in C_0^\infty(\Omega)$ and $\|f_j\|_p = 1$ for all j. By Proposition 3.8 there exists $C > 0$ such that for all $j \in \mathbb{N}$,

$$\int_{\mathbb{R}^n} |f_j(x+h) - f(x)|^p \, dx \leq C |h|^{sp} \to 0 \text{ as } |h| \to 0.$$

Hence by the Riesz–Fréchet–Kolmogorov theorem, there is a subsequence of $\{f_j\}_{j \in \mathbb{N}}$ that converges in $L_p(\Omega)$, to f, say; plainly $\|f\|_{p,\Omega} = 1$. Since X is reflexive, there is a further subsequence $\{g_j\}_{j \in \mathbb{N}}$ that converges weakly in X, to g, say; as this subsequence also converges weakly in $L_p(\Omega)$ we see that $g = f$. As

$$\|f|X\| \leq \lim_{j \to \infty} \inf \|g_j|X\| \,,$$

it follows that $\|f|X\| = \lambda_1$, showing that the infimum is attained.

It remains to deal with the Euler–Lagrange equation. Let u be a minimising function and consider the competing functions

$$v_t(x) := u(x) + t\phi(x), \ \phi \in C_0^\infty(\Omega), t \in \mathbb{R}.$$

Because we have a minimum we must have

$$\frac{d}{dt} \left\{ \frac{\int_{\mathbb{R}^n} \int_{\mathbb{R}^n} \frac{|v_t(y) - v_t(x)|^p}{|y-x|^{n+sp}} \, dx \, dy}{\int_{\mathbb{R}^n} |v_t(x)|^p \, dx} \right\} = 0 \text{ at } t = 0.$$

This immediately gives the Euler–Lagrange equation.   □

Remark 4.2

Note that since the inequality

$$||u(x)| - |u(y)|| \leq |u(x) - u(y)|$$

is strict at almost all points x, y such that $u(x)u(y) < 0$, no minimiser can change sign. It should also be observed that $\lambda_1(p, s, \Omega)$ is the reciprocal of the best constant in the (s, p)-Friedrichs inequality; in fact,

$$\lambda_1(p, s, \Omega) \geq 1/C(n, s, p, \Omega) \,,$$

where $C(n, s, p, \Omega)$ is given in Proposition 3.5.

As in Section 3.4, given $\lambda \in \mathbb{R}$ we say that $u \in X \backslash \{0\}$ is a weak solution of the eigenvalue problem

$$(-\Delta)_p^s u = \lambda |u|^{p-2} u \text{ in } \Omega, u = 0 \text{ in } \mathbb{R}^n \backslash \Omega, \qquad (4.1.3)$$

if

$$\int_{\mathbb{R}^n} \int_{\mathbb{R}^n} \frac{|u(x)-u(y)|^{p-2} (u(x)-u(y)) (\phi(x)-\phi(y))}{|x-y|^{n+sp}} \, dx \, dy = \lambda \int_{\mathbb{R}^n} |u|^{p-2} u\phi \, dx$$
(4.1.4)

for all $\phi \in X$; if such a function u exists, the corresponding λ is an *eigenvalue* and u is a λ-*eigenfunction*.

Every solution of the Euler–Lagrange equation is bounded. When $sp > n$ the Euler–Lagrange equation is not needed: in fact since, by Proposition 3.8, X is embedded in $C^\alpha(\mathbb{R}^n)$, where $\alpha = s - n/p$, the boundedness is clear. Much more effort is needed when $sp \le n$.

Theorem 4.3 *Suppose that $sp \le n$ and let u be a minimiser of the Rayleigh quotient. Then $u \in L_\infty(\mathbb{R}^n)$; and if $sp < n$,*

$$\|u\|_\infty \le C(n, p, s)\lambda_1^{n/\left(sp^2\right)} \|u\|_p.$$

For a proof of this result we refer to [28], Theorem 3.3 and [84], Theorem 3.2.

Corollary 4.4 *Let $s \in (0, 1)$, $p \in (1, \infty)$. Then every (s, p)-eigenfunction is continuous.*

Proof If $sp > n$ there is nothing to prove because of Proposition 3.8. If $sp \le n$, we know from Theorem 4.3 that eigenfunctions are bounded. The continuity then follows from Theorem 1.5 of [111] (see Corollary 3.14 of [29]). □

We now turn to further basic properties of eigenfunctions. The next theorem was established in [27] when Ω is connected, the general case being proved in [29]. A crucial step in the argument is the following logarithmic estimate, given in [50]. It concerns functions $u \in X = \overset{0}{X}{}_p^s(\Omega)$ that are supersolutions (of the problem $(-\Delta)^s u = 0$ in Ω, $u = 0$ in $\mathbb{R}^n \backslash \Omega$) in the sense that

$$\int_{\mathbb{R}^n} \int_{\mathbb{R}^n} \frac{|u(x) - u(y)|^{p-2} (u(x) - u(y))}{|x - y|^{n+sp}} (\phi(x) - \phi(y)) \, dx \, dy \ge 0$$

for all $\phi \in C_0^\infty(\Omega)$, $\phi \ge 0$.

Lemma 4.5 *Let $s \in (0, 1)$, $p \in (1, \infty)$, and let $u \in X = \overset{0}{X}{}_p^s(\Omega)$ be a supersolution such that $u \ge 0$ in $B(x_0, 2r)$ for some $r > 0$ and $x_0 \in \Omega$ with $\overline{B(x_0, 2r)} \subset \Omega$. Then there is a constant $C = C(n, p, s)$ such that for all $\delta > 0$,*

$$\int_{B(x_0,r)} \int_{B(x_0,r)} \left| \log \left(\frac{u(x)+\delta}{u(y)+\delta} \right) \right|^p \frac{1}{|x-y|^{n+sp}} \, dx \, dy$$

$$\leq Cr^{n-sp} \left\{ \delta^{1-p} r^{sp} \int_{\mathbb{R}^n \setminus B(x_0,2r)} \frac{u_-(y)^{p-1}}{|y-x_0|^{n-sp}} \, dy + 1 \right\},$$

where $u_- = \max\{-u, 0\}$.

Theorem 4.6 *Let* $u \in X = \overset{0}{X}{}^s_p(\Omega)$ *be a non-negative* (s, p)-*eigenvector with corresponding eigenvalue* λ. *Then* $u > 0$ *a.e. in* Ω.

Proof First assume that Ω is connected and let K be a compact connected subset of Ω, so that $K \subset \{x \in \Omega : \text{dist}(x, \partial\Omega) > 2r\}$ for some $r > 0$. Then K can be covered by balls $B(x_i, r/2)$ $(i = 1, ..., k)$ with each $x_i \in K$ and

$$|B(x_i, r/2) \cap B(x_{i+1}, r/2)| > 0 \ (i = 1, ..., k-1). \tag{4.1.5}$$

Suppose that $u = 0$ on a subset of K with positive measure. Then there exists $i \in \{1, ..., k-1\}$ such that

$$Z := \{x \in B(x_i, r/2) : u(x) = 0\}$$

has positive measure. For each $\delta > 0$ set

$$F_\delta(x) = \log \left(1 + \frac{u(x)}{\delta} \right), \ x \in B(x_i, r/2).$$

Then $F_\delta(x) = 0$ for all $x \in Z$, and so for all $x \in B(x_i, r/2)$ and $y \in Z \setminus \{x\}$,

$$|F_\delta(x)|^p = \frac{|F_\delta(x) - F_\delta(y)|^p}{|x-y|^{n+sp}} |x-y|^{n+sp},$$

from which we have, on integrating with respect to $y \in Z$ and $x \in B(x_i, r/2)$,

$$\int_{B(x_i,r/2)} |F_\delta(x)|^p \, dx \leq \frac{r^{n+sp}}{|Z|} \int_{B(x_i,r/2)} \int_{B(x_i,r/2)} \frac{|F_\delta(x) - F_\delta(y)|^p}{|x-y|^{n+sp}} \, dx \, dy. \tag{4.1.6}$$

Together with Lemma 4.5, noting that $u_- = 0$, this shows that

$$\int_{B(x_i,r/2)} \left| \log \left(1 + \frac{u(x)}{\delta} \right) \right|^p dx \leq Cr^{2n}/|Z|,$$

where C is independent of δ. As this holds for arbitrarily small $\delta > 0$, it follows that $u = 0$ a.e. in $B(x_i, r/2)$. In view of (4.1.5) this argument can be repeated for $B(x_{i\pm 1}, r/2)$, from which we see that $u = 0$ a.e. in K. This contradicts our original assumption and we conclude that $u > 0$ a.e. on K.

Now recall that u is an eigenvector and so is not identically zero in Ω. Since Ω is connected, there is a sequence $\{K_m\}$ of connected compact subsets of Ω such that for each m, $|\Omega \setminus K_m| < 1/m$ and u is not identically zero in K_m. By what we have proved, $u > 0$ a.e. on each K_m. It follows that $u > 0$ a.e. on Ω.

To complete the proof it remains to deal with the case in which Ω is not connected. We know that $u > 0$ a.e. on each connected component of Ω on which it is not identically zero. Put $\Omega_1 = \{x \in \Omega : u(x) > 0\}$ and suppose there is a connected component Ω_2 of Ω on which u is identically zero. Let $\phi \in C_0^\infty(\Omega_2)$ be a non-negative test function that is not identically zero. Then

$$0 = \lambda \int_\Omega u^{p-1} \phi \, dx = \int_{\mathbb{R}^n} \int_{\mathbb{R}^n} \frac{|u(x) - u(y)|^{p-2} (u(x) - u(y))}{|x - y|^{n+sp}} (\phi(x) - \phi(y)) \, dx \, dy$$

$$= -2 \int_{\Omega_1} \int_{\Omega_2} \frac{u(x)^{p-1}}{|x - y|^{n+sp}} \phi(y) \, dx \, dy.$$

Thus u is identically zero in Ω_1 and we have a contradiction that establishes the theorem. $\qquad\square$

All eigenfunctions except those corresponding to λ_1 change sign. Formally,

Theorem 4.7 *Let $v \in X = \overset{0}{X}_p^s(\Omega)$ be a solution of (4.1.4), with corresponding eigenvalue λ, such that $v > 0$ in Ω. Then $\lambda = \lambda_1 (p, s, \Omega)$.*

The ingenious proof is given in [84], Theorem 4.1. Theorem 4.2 of the same paper shows that λ_1 is simple:

Theorem 4.8 *Let $s \in (0, 1)$ and $p \in (1, \infty)$. All positive eigenfunctions corresponding to $\lambda_1 (p, s, \Omega)$ are proportional.*

For later convenience we now summarise these properties of the first eigenvalue.

Theorem 4.9 *Let $s \in (0, 1)$ and $p \in (1, \infty)$; suppose that Ω is bounded. Then*

 (i) *any first (s, p)-eigenfunction must be strictly positive (or strictly negative);*
 (ii) *$\lambda_1 (p, s, \Omega)$ is simple;*
 (iii) *any eigenfunction corresponding to an eigenvector $\lambda > \lambda_1 (p, s, \Omega)$ must change sign.*

4.2 The Spectrum

In contrast to the standard terminology used for linear operators, the set of all eigenvalues of the eigenvalue problem (4.1.3) is called the *spectrum* of (4.1.3) and is denoted by $\sigma(s, p)$; given $\lambda \in \sigma(s, p)$, the set of all λ-eigenfunctions is the *λ-eigenspace*.

Proposition 4.10 *The spectrum $\sigma(s, p)$ is closed.*

Proof   Let $\lambda \in \overline{\sigma(s, p)}$; then there is a sequence $\{\lambda_k\}$ of eigenvalues such that $\lambda_k \to \lambda$. Put $X = \overset{0}{X_p^s}(\Omega)$ and define $A \colon X \to X^*$ as in 3.4, so that for all $u, v \in X$,

$$\langle Au, v \rangle = \int_{\mathbb{R}^n} \int_{\mathbb{R}^n} \frac{|u(x) - u(y)|^{p-2} \, (u(x) - u(y)) \, (v(x) - v(y))}{|x - y|^{n+sp}} \, dx \, dy,$$

and

$$\lambda_k = [u_k]_{s,p,\mathbb{R}^n}^p = \|u_k | X\|^p \text{ for some } u_k \in X \text{ with } \|u_k\|_p = 1 \; (k \in \mathbb{N}).$$

Since $\{u_k\}$ is bounded in the reflexive space X, there is a subsequence, still denoted by u_k for convenience, and an element u of X, such that $u_k \rightharpoonup u$ in X; as X is compactly embedded in $L_p(\Omega)$, we may and shall suppose that $u_k \to u$ in $L_p(\Omega)$. Moreover,

$$|\langle Au_k, u_k - u \rangle| = \lambda_k \left| \int_{\mathbb{R}^n} |u_k|^{p-2} u_k (u_k - u) \, dx \right|$$

$$\leq \lambda_k \|u_k - u\|_p \|u_k\|_p^{p/p'} \to 0$$

as $k \to \infty$. Thus

$$\lim_{k \to \infty} |\langle Au_k - Au, u_k - u \rangle| = 0,$$

and as by Lemma 3.31 the map A is of type $(S)_+$, it follows that $u_k \to u$ in X. By Proposition 1.1.26 of [61], applied to the duality map A and the uniformly smooth space X^*, we see that $Au_k \to Au$ in X^*. Let $J \colon L_p(\Omega) \to L_{p'}(\Omega)$ be the duality map with gauge function $t \longmapsto t^{p-1}$, so that $Jf = |f|^{p-2} f$ $(f \in L_p(\Omega))$. Then for all $v \in X$,

$$\langle Au, v \rangle = \lim_{k \to \infty} \langle Au_k, v \rangle = \lim_{k \to \infty} \lambda_k \int_{\mathbb{R}^n} |u_k|^{p-2} u_k v \, dx = \lim_{k \to \infty} \lambda_k \langle Ju_k, v \rangle$$

$$= \lambda \langle Ju, v \rangle = \lambda \int_{\mathbb{R}^n} |u|^{p-2} uv \, dx,$$

and $\|u\|_{p,\Omega} = 1$, so that $\lambda \in \sigma(s, p)$. $\square$

Theorem 4.11   *If* $\{\Omega_j\}$ *is a non-decreasing sequence of domains such that* $\Omega = \bigcup_{j=1}^{\infty} \Omega_j$, *then* $\lambda_1(p, s, \Omega_j) \downarrow \lambda_1(p, s, \Omega)$.

Proof   Evidently $\lambda_1(p, s, \Omega_j)$ decreases as j increases; hence the limit exists. Given $\varepsilon > 0$, there is a function $\phi \in C_0^{\infty}(\Omega)$ such that

$$[\phi]_{s,p,\mathbb{R}^n}^p / \|\phi\|_{p,\mathbb{R}^n}^p < \lambda_1(p, s, \Omega) + \epsilon, \tag{4.2.1}$$

since $\lambda_1(p, s, \Omega)$ is the infimum. However, as supp $\phi \subset \Omega_j$ for large enough j, the function ϕ can be used as a test function in the Rayleigh quotient for Ω_j, and so

$$\lambda_1\left(p, s, \Omega_j\right) < \lambda_1\left(p, s, \Omega\right) + \epsilon$$

for all large enough j. The result follows. $\qquad\qquad\qquad\qquad\qquad$ □

Let $S_p(\Omega) = \left\{u \in X \colon \|u\|_p = 1\right\}$ and define

$$\lambda_2(s, p, \Omega) = \inf_{f \in \mathcal{C}_1(\Omega)} \max_{u \in im(f)} \|u\|_{X_p^s(\Omega)}^p,$$

where

$$\mathcal{C}_1(\Omega) = \left\{f \colon \mathcal{S}^1 \to S_p(\Omega) \colon f \text{ odd and continuous}\right\}.$$

Theorem 4.12 *Let $s \in (0, 1)$ and $p \in (1, \infty)$; suppose that Ω is bounded. Then $\lambda_2(s, p, \Omega)$ is an (s, p)-eigenvalue, $\lambda_2(s, p, \Omega) > \lambda_1(s, p, \Omega)$, and for every (s, p)-eigenvalue $\lambda > \lambda_1(p, s, \Omega)$ we have $\lambda \geq \lambda_2(s, p, \Omega)$.*

This is given in detail in [29], Theorem 4.1 and Proposition 4.2. Here we simply sketch some of the main ideas used to establish the result.

To prove that $\lambda_2(s, p, \Omega)$ is an (s, p)-eigenvalue amounts to showing that it is a critical point of the functional

$$\Phi_{s,p}(u) := \int_{\mathbb{R}^n} \int_{\mathbb{R}^n} \frac{|u(x) - u(y)|^p}{|x - y|^{n+sp}} \, dx \, dy$$

defined on the manifold

$$S_p(\Omega) := \left\{u \in X(\Omega) \colon \|u\|_{p,\Omega} = 1\right\}.$$

The strategy is to show that $\Phi_{s,p}$ satisfies the Palais–Smale condition: once this is done, the result will follow from Theorem 1.8. To do this, let $\{u_k\}_{k \in \mathbb{N}}$ be a sequence in $S_p(\Omega)$ such that there exists $C > 0$ with

$$\Phi_{s,p}(u_k) \leq C \ (k \in \mathbb{N}) \text{ and } \lim_{k \to \infty} \left\|\Phi_{s,p}'(u_k)|T_{u_k}S_p(\Omega)\right\| = 0. \qquad (4.2.2)$$

In this context the tangent space to $S_p(\Omega)$ at u_k is given by

$$T_{u_k}S_p(\Omega) = \left\{\phi \in X \colon \int_\Omega |u_k|^{p-2} u_k \phi \, dx = 0\right\}.$$

By the second part of (4.2.2), there exists $\{\varepsilon_k\}_{k \in \mathbb{N}}$, $\varepsilon_k > 0$, $\varepsilon_k \to 0$ such that for all $k \in \mathbb{N}$,

$$\left|\Phi_{s,p}'(u_k)(\phi)\right| \leq \varepsilon_k \|\phi\|_X \text{ for all } \sigma \in T_{u_k}S_p(\Omega).$$

By the first part of (4.2.2), there is a subsequence of $\{u_k\}_{k \in \mathbb{N}}$, still denoted by $\{u_k\}_{k \in \mathbb{N}}$ for convenience, and a function $u \in X$ such that $u_k \to u$ in $L_p(\Omega)$ and $u_k \rightharpoonup u$ in X. Clearly $u \in S_p(\Omega)$. It remains to prove that a further subsequence of $\{u_k\}_{k \in \mathbb{N}}$ converges to u in X: details of the technical argument needed to establish this are given in [29].

To prove that $\lambda_2(s, p, \Omega) > \lambda_1(s, p, \Omega)$, suppose that this is false, so that

$$\lambda_2(s, p, \Omega) = \inf_{f \in \mathcal{C}_1(\Omega)} \max_{u \in im(f)} \|u\|_X = \lambda_1(s, p, \Omega).$$

Hence given any $k \in \mathbb{N}$, there is an odd continuous map $f_k \colon \mathbb{S}^1 \to S_p(\Omega)$ such that

$$\max_{u \in f_k(\mathbb{S}^1)} \|u\|_X \le \lambda_1(s, p, \Omega) + k^{-1}.$$

Let $u_1 \in S_p(\Omega)$ be the unique (modulo the choice of sign) global minimiser, and for small enough positive ε consider the disjoint neighbourhoods

$$U_+ := \left\{ u \in S_p(\Omega) : \|u - u_1\|_p < \varepsilon, \right\},$$
$$U_- := \left\{ u \in S_p(\Omega) : \|u - (-u_1)\|_p < \varepsilon, \right\};$$

note that $U_+ \cup U_-$ is symmetric and disconnected. For every $k \in \mathbb{N}$, the image $f_k(\mathbb{S}^1)$ of $\mathbb{S}^1$ under the odd continuous map f_k is symmetric and connected: it follows that there exists $v_k \in f_k(\mathbb{S}^1) \setminus (U_+ \cup U_-)$. The sequence $\{v_k\}_{k \in \mathbb{N}}$ is contained in $S_p(\Omega)$ and is bounded in X: by passage to a subsequence if necessary we see that there exists $v \in S_p(\Omega)$ such that $v_k \rightharpoonup v$ in X and $v_k \to v$ in $L_p(\Omega)$. Hence

$$\|v\|_X \le \liminf_{k \to \infty} \|v_k\|_X = \lambda_1(s, p, \Omega).$$

Thus $v \in S_p(\Omega)$ is a global minimiser, so that either $v = u_1$ or $v = -u_1$. But $v \in S_p(\Omega) \setminus (U_+ \cup U_-)$ and we have a contradiction. Hence $\lambda_2(s, p, \Omega) > \lambda_1(s, p, \Omega)$.

For the ingenious and technical proof of the last part of the theorem we refer to Proposition 4.2 of [29].

This result shows that $\lambda_2(s, p, \Omega)$ may properly be called the second (s, p)-eigenvalue. It also shows that $\lambda_1(s, p, \Omega)$ is isolated.

4.3 Inequalities of Faber–Krahn Type

In this section we deal with a fractional version of the celebrated Faber–Krahn inequality concerning the first eigenvalue $\lambda_1(p, \Omega)$ of the Dirichlet p-Laplacian. This asserts that balls minimise the first eigenvalue among open sets with given volume; more precisely,

$$\lambda_1(p, B) \le \lambda_1(p, \Omega),$$

where

$$\lambda_1(p, \Omega) = \min \left\{ \int_\Omega |\nabla u|^p \, dx \colon u \in \overset{0}{W}{}_p^1(\Omega), \ \|u\|_p = 1 \right\}$$

and B is a ball with the same measure as Ω. A crucial component of the proof is the Pólya–Szegő inequality, which involves the notion of the symmetric rearrangement of a function and which we now explain. Given $u: \mathbb{R}^n \to \mathbb{R}^+ \cup \{\infty\}$, its symmetric rearrangement is defined to be the unique function $u^\star: \mathbb{R}^n \to \mathbb{R}^+ \cup \{\infty\}$ such that for all $\lambda \geq 0$, there exists $R \geq 0$ with

$$B_R = \left\{x \in \mathbb{R}^n: u^\star(x) > \lambda\right\} \text{ and } |B_R| = |\{x \in \mathbb{R}^n: u(x) > \lambda\}|.$$

The function $u^\star$ is radial and radially decreasing. It is easy to see that if u belongs to $L_p(\mathbb{R}^n)$ then so does $u^\star$, which has the same L_p norm. The famous Pólya–Szegő inequality asserts that if $u \in W_p^1(\mathbb{R}^n)$ is non-negative, then $u^\star \in W_p^1(\mathbb{R}^n)$ and

$$\int_{\mathbb{R}^n} |\nabla u^\star|^p \, dx \leq \int_{\mathbb{R}^n} |\nabla u|^p \, dx.$$

This was extended to fractional Sobolev spaces in [9], Theorem 9.2 (see also [70], Theorem A1); more precisely,

Theorem 4.13 *Let $s \in (0, 1)$ and $p \in (1, \infty)$. Then for all $u \in W_p^s(\mathbb{R}^n)$,*

$$[u]_{s,p,\mathbb{R}^n} \geq \left[u^\star\right]_{s,p,\mathbb{R}^n}.$$

Armed with this inequality, Brasco, Lindgren and Parini [28] established the following version of the Faber–Krahn inequality.

Theorem 4.14 *Let $s \in (0, 1)$ and $p \in (1, \infty)$; suppose that Ω is bounded. Then for every open ball $B \subset \mathbb{R}^n$,*

$$|\Omega|^{sp/n} \lambda_1(s, p, \Omega) \geq |B|^{sp/n} \lambda_1(s, p, B). \tag{4.3.1}$$

If equality holds, then Ω is a ball.

Proof Note that $\lambda_1(s, p, t\Omega) = t^{-sp}\lambda_1(s, p, \Omega)$. Without loss of generality we may suppose that $|\Omega| = |B|$. Then (4.3.1) follows immediately from Theorem 4.13. As for equality, if $|\Omega| = |B|$ and $\lambda_1(s, p, \Omega) = \lambda_1(s, p, B)$, then by Theorem A.1 of [82], (4.3.1) holds with equality. However, again by Theorem A1 of [82], any first eigenfunction with respect to Ω must coincide with a translate of a radially symmetric decreasing function, which means that Ω must be a ball. $\qquad\square$

A lower bound for the second eigenvalue was obtained in [29]; their proof relies on the following lemma.

Lemma 4.15 *Let $s \in (0, 1)$ and $p \in (1, \infty)$; suppose that Ω is bounded and let λ be an eigenvalue with corresponding eigenvector $u \in S_p(\Omega)$ and with $\lambda > \lambda_1(s, p, \Omega)$. Put*

$$\Omega_+ = \{x \in \Omega : u(x) > 0\}, \ \Omega_- = \{x \in \Omega : u(x) < 0\}.$$

Then

$$\lambda > \max \{\lambda_1(s, p, \Omega_+), \lambda_1(s, p, \Omega_-)\}.$$

Proof By Corollary 4.4, u is continuous in Ω; hence $\Omega_\pm$ are open and $\lambda_1(s, p, \Omega_\pm)$ are well defined. Write $u = u_+ - u_-$, where u_+ and u_- are the positive and negative parts of u; recall that u is sign-changing. Use u_+ as the test function in the equation satisfied again by u: this gives

$$\lambda \int_\Omega |u_+|^p \, dx = \int_{\mathbb{R}^n} \int_{\mathbb{R}^n} \frac{|u(x) - u(y)|^{p-2} (u(x) - u(y))}{|x - y|^{n+sp}} (u_+(x) - u_+(y)) \, dx \, dy.$$

Application of Lemma 1.9 with $a = u_+(x) - u_+(y)$ and $b = u_-(x) - u_-(y)$ gives

$$\lambda \int_\Omega |u_+|^p \, dx > \int_{\mathbb{R}^n} \int_{\mathbb{R}^n} \frac{|u(x) - u(y)|^p}{|x - y|^{n+sp}} \, dx \, dy.$$

As u_+ is admissible for the variational problem defining $\lambda_1(s, p, \Omega_+)$, it follows that $\lambda > \lambda_1(s, p, \Omega_+)$. In the same way, using Lemma 1.9 again, this time with $a = u_-(x) - u_-(y)$ and $b = u_+(x) - u_+(y)$, we find that $\lambda > \lambda_1(s, p, \Omega_-)$, and the proof is complete. $\square$

For the classical Laplacian on Ω it is a familiar fact that the restriction of a higher eigenfunction (with corresponding eigenvalue λ) to one of its nodal domains, Ω_1 say, is a first eigenfunction on Ω_1, with corresponding eigenvalue λ. The inequality of the last lemma illustrates the sharp contrast between the classical result and that of the fractional, nonlocal situation considered here. The following theorem is given in [29].

Theorem 4.16 *Let $s \in (0, 1)$ and $p \in (1, \infty)$; suppose that Ω is bounded. Then for every ball $B \subset \mathbb{R}^n$ with $|B| = |\Omega|/2$,*

$$\lambda_2(s, p, \Omega) > \lambda_1(s, p, B). \tag{4.3.2}$$

Equality is never attained, but the estimate is sharp in the sense that given any sequences $\{x_k\}, \{y_k\}$ in $\mathbb{R}^n$ with $\lim_{k \to \infty} |x_k - y_k| = \infty$, and with $\Omega_k := B(x_k, R) \cup B(y_k, R)$, where $R > 0$, then

$$\lim_{k \to \infty} \lambda_2(s, p, \Omega_k) = \lambda_1(s, p, B_R).$$

Proof Let $u \in S_p(\Omega)$ be an eigenfunction with corresponding eigenvalue $\lambda_2(s, p, \Omega)$; define

$$\Omega_+ = \{x \in \Omega : u(x) > 0\}, \ \Omega_- = \{x \in \Omega : u(x) < 0\}$$

(recall that u is sign-changing). By Lemma 4.15 and Theorem 4.14,

$$\lambda_2(s, p, \Omega) > \lambda_1(s, p, \Omega_+) > \lambda_1(s, p, B_{R_1}), \lambda_2(s, p, \Omega) > \lambda_1(s, p, \Omega_-)$$
$$> \lambda_1(s, p, B_{R_2}),$$

where $|B_{R_1}| = |\Omega_+|$ and $|B_{R_2}| = |\Omega_-|$. Hence

$$\lambda_2(s, p, \Omega) > \max\left\{\lambda_1(s, p, B_{R_1}), \lambda_1(s, p, B_{R_2})\right\}. \tag{4.3.3}$$

The scaling properties of λ_1 imply that

$$\lambda_1(s, p, B_R) = R^{-s/p}\lambda_1(s, p, B_1);$$

also, $|B_{R_1}| + |B_{R_2}| \leq |\Omega|$. As the right-hand side of (4.3.3) is minimised when $|B_{R_1}| = |B_{R_2}| = |\Omega|/2$, (4.3.2) follows.

To complete the proof, define Ω_k as in the statement of the Theorem; we may suppose that $B(x_k, R)$ and $B(y_k, R)$ are disjoint for all large enough k. Let u, v be the positive normalised first eigenvalues on $B(x_k, R)$, $B(y_k, R)$ respectively: their form does not depend on the centre of the ball. Put

$$a(x, y) = u(x) - u(y), b(x, y) = v(x) - v(y).$$

By Lemma 1.10,

$$\lambda_2(s, p, \Omega_k) \leq \max_{|w_1|^p + |w_2|^p = 1} \int_{\mathbb{R}^n} \int_{\mathbb{R}^n} \frac{|w_1 a - w_2 b|^p}{|x - y|^{n+sp}} \, dx \, dy$$

$$\leq \max_{|w_1|^p + |w_2|^p = 1} \left\{ \begin{array}{l} \int_{\mathbb{R}^n} \int_{\mathbb{R}^n} \frac{|w_1|^p |a|^p}{|x-y|^{n+sp}} \, dx \, dy + \int_{\mathbb{R}^n} \int_{\mathbb{R}^n} \frac{|w_2|^p |b|^p}{|x-y|^{n+sp}} \, dx \, dy \\ + c_p \int_{\mathbb{R}^n} \int_{\mathbb{R}^n} \frac{\left(|w_1 a|^2 + |w_2 a|^2\right)^{(p-2)/2} |w_1 w_2 ab|}{|x-y|^{n+sp}} \, dx \, dy \end{array} \right\}$$

$$= \lambda_1(s, p, B_R) + c_p \int_{\mathbb{R}^n} \int_{\mathbb{R}^n} \frac{\left(|w_1 a|^2 + |w_2 a|^2\right)^{(p-2)/2} |w_1 w_2 ab|}{|x - y|^{n+sp}} \, dx \, dy.$$

Since $ab = -u(x)v(y) - u(y)v(x)$, the numerator in the last integral is nonzero only if $(x, y) \in B(x_k, R) \times B(y_k, R)$ or $(x, y) \in B(y_k, R) \times B(x_k, R)$. With

$$C := 2 \max_{|w_1|^p + |w_1|^p = 1} \int_{B(x_k, R)} \int_{B(y_k, R)} \left(|w_1 a|^2 + |w_2 a|^2\right)^{(p-2)/2} |w_1 w_2 ab| \, dx \, dy,$$

we thus have

$$\lim_{k \to \infty} \lambda_2(s, p, \Omega_k) \leq \lambda_1(s, p, B_R) + \lim_{k \to \infty} \frac{c_p C}{(|x_k - y_k| - 2C)^{n+sp}} = \lambda_1(s, p, B_R),$$

which completes the proof. □

Remark 4.17 This is the nonlocal version of the Hong–Krahn–Szegö inequality (see [26], Theorem 3.2) which claims that, among sets of prescribed measure, the second eigenvalue of the Dirichlet Laplacian is minimised when the underlying set is the disjoint union of two equal balls. Note that Theorem 4.16 implies that (in scalar invariant form) for any ball $B \subset \mathbb{R}^n$,

$$\lambda_2(s, p, \Omega) > (2 |B| / |\Omega|)^{sp/n} \lambda_1(s, p, B).$$

A survey of eigenvalue bounds for fractional Laplacians and fractional Schrödinger operators is given in [79]. See also [103] and [137].

5

Classical (Local) Hardy Inequalities

5.1 Inequalities on $\mathbb{R}^n$

In [94], Hardy proved the inequality

$$\int_0^\infty \left(\frac{1}{x} \int_0^x F(t)\,dt \right)^p dx \leq \left(\frac{p}{p-1} \right)^p \int_0^\infty F(x)^p dx \qquad (5.1.1)$$

for non-negative functions F on $[0, \infty)$ with $1 < p < \infty$. Landau in [114] showed that the constant $\left(\frac{p}{p-1} \right)^p$ is sharp and that equality is only possible if $F = 0$. On putting $f(x) = \int_0^x F(t)\,dt$, one obtains the more familiar form

$$\int_0^\infty \frac{f(x)^p}{x^p}\,dx \leq \left(\frac{p}{p-1} \right)^p \int_0^\infty f'(x)^p dx, \qquad (5.1.2)$$

satisfied by functions f with $f' \in L_p(0, \infty)$ and $\lim_{x \to 0+} f(x) = 0$.

The analogue of (5.1.2) in $\mathbb{R}^n$ for $n > 1$ is

$$\int_{\mathbb{R}^n} \frac{|f(x)|^p}{|x|^p}\,dx \leq \left| \frac{p}{p-n} \right|^p \int_{\mathbb{R}^n} |\nabla f(x)|^p dx, \qquad (5.1.3)$$

where $\nabla f(x) = (\partial f / \partial x_1, \ldots, \partial f / \partial x_n)$, the gradient of f, and with $|\nabla f(x)| = \left(\sum_{i=1}^n |\partial f / \partial x_i|^2 \right)^{1/2}$. The inequality (5.1.3) holds for all $f \in C_0^\infty(\mathbb{R}^n \setminus \{0\})$ if $n < p < \infty$ and all $f \in C_0^\infty(\mathbb{R}^n)$ for $1 \leq p < n$; see [15], Section 1.2. Since $C_0^\infty(\mathbb{R}^n \setminus \{0\})$ is dense in $\overset{0}{\mathcal{D}}{}^1_p(\mathbb{R}^n \setminus \{0\})$, (5.1.3) is satisfied for all $f \in \overset{0}{\mathcal{D}}{}^1_p(\mathbb{R}^n \setminus \{0\})$ when $n < p < \infty$ and similarly for all $f \in \overset{0}{\mathcal{D}}{}^1_p(\mathbb{R}^n)$ for $1 \leq p < n$; recall that $\overset{0}{\mathcal{D}}{}^1_p(\Omega)$, is the *homogeneous Sobolev space* defined in Section 2.2, namely, the completion of $C_0^\infty(\Omega)$ with respect to the norm $u \longmapsto \|\nabla u\|_{p,\Omega}$.

In the case $1 \leq p < n$, there is an equivalence between (5.1.3) (for $f \in \overset{0}{\mathcal{D}}{}^1_p(\mathbb{R}^n)$) and an optimal Sobolev inequality

$$\|f\|_{p*,p} \leq S_{n,p}\|\nabla f\|_p, \quad f \in \overset{0}{D^1_p}(\mathbb{R}^n), p^* = np/(n-p), \tag{5.1.4}$$

demonstrated by Alvino in [10]; see also [145]. Alvino's constant

$$S_{n,p} = \frac{p}{n-p} \frac{\left[\Gamma(1+n/2)\right]^{1/n}}{\sqrt{\pi}} = \frac{p}{n-p}\left(\frac{n}{\omega_{n-1}}\right)^{1/n}$$

is best possible and is the norm of the embedding $\overset{0}{\mathcal{D}^1_p}(\mathbb{R}^n) \hookrightarrow L_{p*,p}(\mathbb{R}^n)$ which is optimal in the sense that the target space $L_{p*,p}(\mathbb{R}^n)$ is the smallest among all rearrangement-invariant spaces; see the discussion following Theorem 2.1. The equivalence is observed in [41] to be a consequence of the Pólya–Szegö principle and the Hardy–Littlewood inequality by which the left-hand side of (5.1.3) does not increase under radially decreasing symmetrisation and is equal to the left-hand side of (5.1.4) when f is radially decreasing. Alvino actually proved the more general inequality

$$\|f\|_{p*,p} \leq S_{n,p}\|\nabla f\|_{p,q}, \quad 1 \leq p < n, \quad 1 \leq q \leq p$$

and this was extended to the full range $1 \leq q \leq \infty$ in [41]. Let $\overset{0}{\mathcal{D}^1_{p,q}}(\mathbb{R}^n)$ denote the completion of $C^\infty_0(\mathbb{R}^n)$ with respect to the norm $u \mapsto \|\nabla u\|_{p,q}$. The embedding $\overset{0}{\mathcal{D}^1_{p,q}}(\mathbb{R}^n) \hookrightarrow L_{p*,p}(\mathbb{R}^n)$ is well known in the interpolation theory literature and direct proofs may be found in [11] and [165]. It is then established in [41] that for $1 \leq p < n$, (5.1.3) holds for $f \in \overset{0}{\mathcal{D}^1_p}(\mathbb{R}^n)$ if and only if the Sobolev–Marcinkiewicz embedding inequality

$$\|f\|_{p*,\infty} \leq S_{n,p}\|\nabla f\|_{p,\infty}, \quad S_{n,p} = \frac{p}{n-p}\left(\frac{n}{\omega_{n-1}}\right)^{1/n} \tag{5.1.5}$$

holds for every $f \in D^1 L_{p,\infty}(\mathbb{R}^n) := \{f \in L_{p,\infty}(\mathbb{R}^n) : \|\nabla f\|_{p,\infty} < \infty\}$. In contrast to the Hardy inequality, the best possible constant $S_{n,p}$ in (5.1.5) is attained, an extremal function in $D^1 L_{p,\infty}(\mathbb{R}^n)$ being given by

$$\psi(x) = |x|^{-\frac{n-p}{p}}.$$

The Marcinkiewicz space $L_{p*,\infty}(\mathbb{R}^n)$ (also called the *weak-L_{p*}* space) is the smallest rearrangement-invariant space containing ψ.

The (normalised) distance function

$$d_{p*,\infty}(f) := \inf_{a\in\mathbb{R}} \frac{\|f - a\psi\|_{L_{p*,\infty}(\mathbb{R}^n)}}{\|f\|_{L_{p*,p}(\mathbb{R}^n)}}$$

is defined in [45] and there it is shown that for $n \geq 2$ and $1 < p < n$, there exist constants $C = C(n,p)$ and $\alpha = \alpha(n,p)$ such that

$$\left[1 + Cd_{p*,\infty}(f)^\alpha\right] \int_{\mathbb{R}^n} \frac{|f(x)|^p}{|x|^p}\, dx \le \int_{\mathbb{R}^n} |\nabla f(x)|^p\, dx. \tag{5.1.6}$$

However, while the Hardy inequality (5.1.3) holds for $p = 1$, the inequality (5.1.6) does not; indeed, for $p = 1$, any spherically symmetric function attains equality in (5.1.3).

There is no valid inequality (5.1.3) for $n = p$, see [15], Section 1.2.5. In the case $n = p = 2$ it is proved in [3], Theorem 4.6, that for all $f \in C_0^\infty(\mathbb{R}^2 \setminus \{0\})$ satisfying $\int_{1<|x|<2} f(x)\, dx = 0$, there exists a positive constant C such that

$$\int_{\mathbb{R}^2} \frac{|f(x)|^2}{|x|^2(1 + \log^2 |x|)}\, dx < C \int_{\mathbb{R}^2} |\nabla f(x)|^2 dx. \tag{5.1.7}$$

Solomyak had shown earlier in [163] that the logarithmic factor in (5.1.7) is only needed for radial functions and can be removed for functions satisfying

$$\int_{|x|=R} f(x)\, dx = 0$$

for all $R > 0$. This condition is also imposed in the following weighted inequality of Dubinskii from [53], Theorem 2.1, which covers the case $n = p$:

Theorem 5.1 *Let* $n \ge 2$, $p > 1$ *and suppose that* $u \in L_{p,loc}(\mathbb{R}^n \setminus \{0\})$, $\int_{\mathbb{R}^n} |\nabla u(x)|^p |x|^{(p-n)} dx < \infty$ *and* $\int_{|x|=R} u(x)\, dx = 0$ *for all* $R > 0$. *Then there exists a constant* M *which depends only on* n *and* p *such that*

$$\int_{\mathbb{R}^n} \mu_R(|x|)|u(x)|^p dx \le M \int_{\mathbb{R}^n} |\nabla u(x)|^p |x|^{p-n} dx, \tag{5.1.8}$$

where for $r > 0$,

$$\mu_R(r) = \min\left\{ \frac{1}{r^n \left|\ln\left(\frac{r}{R}\right)\right|^p}, \frac{1}{r^n} \right\}.$$

An interesting counterpart of (5.1.3) on $\mathbb{R}^2 \setminus \{0\}$ was established in [116] by replacing the gradient ∇ with the magnetic gradient $\nabla + i\mathbf{A}$, where $\mathbf{A}$ is a magnetic potential of Aharonov–Bohm type given in polar co-ordinates $x = (r\cos\theta, r\sin\theta)$ by

$$\mathbf{A}(x) = \frac{\psi(\theta)}{r}(-\sin\theta, \cos\theta),$$

where $\psi \in L^\infty(0, 2\pi)$ and

$$\Psi := \frac{1}{2\pi} \int_0^{2\pi} \psi(\theta) d\theta$$

is the magnetic flux; significant features are that the domain $\mathbb{R}^2 \setminus \{0\}$ is not simply connected and the magnetic field curl $\mathbf{A}(x) = 0$ in $\mathbb{R}^2 \setminus \{0\}$. The resulting Laptev–Weidl inequality is that, for all non-trivial $f \in C_0^\infty(\mathbb{R}^2 \setminus \{0\})$,

$$\int_{\mathbb{R}^2} |(\nabla + i\mathbf{A})f(x)|^2 \, dx > \min_{k \in \mathbb{Z}} |k - \Psi|^2 \int_{\mathbb{R}^2} \frac{|f(x)|^2}{|x|^2} \, dx, \qquad (5.1.9)$$

with sharp constant $\min_{k \in \mathbb{Z}} |k - \Psi|^2$. If the magnetic flux Ψ is an integer, the magnetic Laplace operator $-(\nabla + i\mathbf{A})^2$ is unitarily equivalent in $L_2(\mathbb{R}^2)$ to $-\Delta$ and hence there is no non-trivial Hardy inequality. We shall return to this inequality and discrete versions in Section 5.6.

Our main concern in subsequent sections of this chapter will be with the validity and refinements of a general inequality

$$\int_{\Omega} \frac{|f(x)|^p}{\delta(x)^p} \, dx \le C(p, \Omega) \int_{\Omega} |\nabla f(x)|^p dx \qquad (5.1.10)$$

for $f \in C_0^\infty(\Omega)$ and Ω an open connected set (domain) in $\mathbb{R}^n$ with non-empty boundary; in (5.1.10), $\delta(x) = \inf\{|x - y| : y \notin \Omega\}$, the distance of x from the boundary of Ω. Since $C_0^\infty(\Omega)$ is dense in $\overset{0}{W_p^1}(\Omega)$, it would follow that (5.1.10) holds on $\overset{0}{W_p^1}(\Omega)$, and indeed on the larger space $\overset{0}{\mathcal{D}_p^1}(\Omega)$.

The production of papers on the Hardy inequality has mushroomed in this century and significant works continue to appear at an accelerating rate. The selection of results deemed to be of particular significance is inevitably personal and some worthy contributions are bound to be omitted. We make an attempt at a comprehensive coverage within these obvious bounds. Some results are stated without proof, but with what we hope is adequate background information and precise references.

In the range $1 < p \le n$, (5.1.10) was proved in [122] to be valid if $\mathbb{R}^n \setminus \Omega$ is *uniformly p-fat*, and valid for $p = n$ if and only if $\mathbb{R}^n \setminus \Omega$ is *uniformly p-fat*. We refer to [122] and [15] for a definition of the *uniformly p-fat property*, but the following examples may help to put it in perspective:

1. A closed set satisfying the interior cone condition is uniformly p-fat for every $p \in (1, \infty)$.
2. The complement of a Lipschitz domain is uniformly p-fat for every $p \in (1, \infty)$. Recall that a domain is Lipschitz if it is a rotation of a set of the form

$$\{x = (x', x_n) = (x_1, ..., x_{n-1}, x_n) \in \mathbb{R}^n : x_n = \Phi(x')\}$$

where $\Phi \colon \mathbb{R}^{n-1} \to \mathbb{R}$ is a Lipschitz function.

The uniform p-fat property of $\mathbb{R}^n \setminus \Omega$ was proved by Lehrbäck in [120] and [121] to be equivalent to the *pointwise q-Hardy inequality for some* $q \in (1, p)$; this notion was introduced by Hajłasz in [93] and is that there exists a positive constant $c(n, q)$, depending only on n and q, such that for all $f \in C_0^\infty(\Omega)$ (extended by zero to all of $\mathbb{R}^n$),

$$\frac{|f(x)|}{\delta(x)} \le c(n,q) \left[\mathrm{M}\left(|\nabla f|^q(x)\right) \right]^{1/q},$$

where $\mathrm{M}f$ is the maximal function defined for $f \in L_{1,loc}(\mathbb{R}^n)$ by

$$\mathrm{M}f(x) := \sup_{r>0} \frac{1}{|B(x,r)|} \int_{B(x,r)} |f(y)|\, dy.$$

In the range $n < p < \infty$, (5.1.10) was proved in [122] to be valid for all proper open subsets Ω of $\mathbb{R}^n (n \ge 2)$. The weighted inequality in the following theorem includes the case $n < p < \infty$ of [122] and gives the best possible value for the constant $C(p, \Omega)$. It was first proved by Avkhadiev in [7] but alternative proofs have since been given by Chen in [43] and Pinchover and Goel in [148]. The following proof is that in [43].

Theorem 5.2 *Let* $\Omega \subsetneq \mathbb{R}^n, n \ge 2$, *be an arbitrary domain,* $1 < p < \infty$ *and* $\alpha + p > n$. *Then for all f such that* $|f| \in C_0^\infty(\Omega)$,

$$\left(\frac{\alpha+p-n}{p} \right)^p \int_\Omega \frac{|f(x)|^p}{\delta(x)^{p+\alpha}}\, dx \le \int_\Omega \frac{|\nabla f(x)|^p}{\delta(x)^\alpha}\, dx, \qquad (5.1.11)$$

where the constant is sharp.

Hence, in particular, when $n < p < \infty$, *for all* $f \in \overset{0}{D_p^1}(\Omega)$,

$$\left(\frac{p-n}{p} \right)^p \int_\Omega \frac{|f(x)|^p}{\delta^p(x)}\, dx \le \int_\Omega |\nabla f(x)|^p\, dx, \qquad (5.1.12)$$

where the constant is sharp.

Proof An important first step is to show that for any $\beta \ge 2$,

$$\Delta \delta^{2-\beta} \ge (\beta - 2)(\beta - n)\delta^{-\beta} \qquad (5.1.13)$$

in the sense of distributions in Ω, i.e., for every non-negative $\phi \in C_0^\infty(\Omega)$,

$$\int_\Omega \left\{ \delta^{2-\beta} \Delta \phi - (\beta-2)(\beta-n)\delta^{-\beta}\phi \right\} dx \ge 0.$$

To prove this, first observe that for every $a \in \mathbb{R}^n$,

$$u_a(x) := -|x-a|^2 + |x|^2$$

is harmonic in $\mathbb{R}^n$, and that

$$\delta(x) = \min_{a \in \partial\Omega}\{|x-a|\}.$$

Thus

$$-\delta^2(x) + |x|^2 = \max_{a \in \partial\Omega}\{u_a(x)\}$$

and for $0 \le \phi \in C_0^\infty(\Omega)$,

$$\int_\Omega \delta(x)^2 \Delta\phi \, dx = \int_\Omega \left(|x|^2 - \max_{a\in\partial\Omega} u_a(x) \right) \Delta\phi \, dx$$

$$\leq \int_\Omega \Delta \left(|x|^2 - u_a(x) \right) \phi \, dx$$

$$= 2n \int_\Omega \phi \, dx.$$

This means that $-\Delta\delta^2 \geq -2n$ in the distributional sense and setting $\chi(t) := t^{1-\beta/2}$, we have

$$\Delta\chi(\delta^2) = \chi''(\delta^2)|\nabla\delta^2|^2 + \chi'(\delta^2)\Delta\delta^2$$

$$= \beta(\beta - 2)\delta^{-\beta}|\nabla\delta|^2 + (1 - \beta/2)\delta^{-\beta}\Delta\delta^2$$

$$\geq (\beta - 2)(\beta - n)\delta^{-\beta},$$

since $|\nabla\delta| = 1$ a.e. on Ω; this will be proved in Section 5.2 and is a consequence of δ being uniformly Lipschitz and $|\delta(x) - \delta(y)| \leq |x - y|$ for $x, y \in \Omega$. Therefore (5.1.13) is proved.

Hence, for $|f| \in C_0^\infty(\Omega)$,

$$(\beta - 2)(\beta - n) \int_\Omega \frac{|f|^p}{\delta^\beta} dx \leq \int_\Omega \Delta \left(\delta^{2-\beta} \right) |f|^p \, dx$$

$$= \int_\Omega \delta^{2-\beta} \Delta \left(|f|^p \right)$$

$$= - \int_\Omega \nabla \left(\delta^{2-\beta} \right) \cdot \nabla \left(|f|^p \right) dx$$

$$= p(\beta - 2) \int_\Omega \delta^{1-\beta}|f|^{p-1}\nabla\delta \cdot \nabla|f| \, dx,$$

and for $\beta > 2$,

$$\frac{(\beta - n)}{p} \int_\Omega \frac{|f|^p}{\delta^\beta} dx \leq \int_\Omega \delta^{1-\beta}|f|^{p-1}\nabla\delta \cdot \nabla|f| \, dx$$

$$\leq \left(\int_\Omega |f|^p |\nabla\delta|^{\frac{p}{p-1}} \delta^{-\beta} dx \right)^{\frac{p-1}{p}} \left(\int_\Omega |\nabla|f||^p \delta^{p-\beta} \right)^{\frac{1}{p}}$$

$$\leq \left(\int_\Omega |f|^p \delta^{-\beta} dx \right)^{\frac{p-1}{p}} \left(\int_\Omega |\nabla f|^p \delta^{p-\beta} \right)^{\frac{1}{p}},$$

since $|\nabla|f|| \leq |\nabla f|$ a.e. The inequality (5.1.11) follows on putting $\beta = p + \alpha$.

To prove that the constant is sharp, Chen considers

$$\Omega = B_2 := \{x: 0 < |x| < 2\},$$

and sets $\gamma_\varepsilon = (\alpha + p - n)/p + \varepsilon$, $\varepsilon > 0$. Let f_ε be a test function with compact support in B_2 and such that $f_\varepsilon(x) = |x|^{\gamma_\varepsilon}$ on $B_1 = \{x: 0 < |x| < 1\}$. On using polar co-ordinates, we obtain for small $\varepsilon > 0$,

$$\int_{B_2} \frac{|f_\varepsilon|^p}{\delta^{p+\alpha}} \, dx = \int_{B_1} \frac{|f_\varepsilon|^p}{\delta^{p+\alpha}} \, dx + O(1)$$

$$= \omega_n \int_0^1 r^{-1+p\varepsilon} \, dr + O(1)$$

$$= \omega_n (p\varepsilon)^{-1} + O(1).$$

Similarly, we have

$$\int_{B_2} \frac{|\nabla f_\varepsilon|^p}{\delta^\alpha} \, dx = \omega_n \gamma_\varepsilon^p \int_0^1 r^{p(\gamma_\varepsilon - 1) - \alpha + n - 1} \, dr + O(1)$$

$$= \omega_n \gamma_\varepsilon^p (p\varepsilon)^{-1} + O(1).$$

Thus

$$\lim_{\varepsilon \to 0+} \frac{\int_{B_2} \frac{|\nabla f_\varepsilon|^p}{\delta^\alpha} \, dx}{\int_{B_2} \frac{|f_\varepsilon|^p}{\delta^{p+\alpha}} \, dx} = \left(\frac{\alpha + p - n}{p} \right)^p. \tag{5.1.14}$$

Since every f_ε may be approximated by functions in $C_0^\infty(B_2)$ with respect to the norm

$$\left(\int_{B_2} |\cdot|^p / \delta^{p+\alpha} \, dx \right)^{1/p} + \left(\int_{B_2} |\nabla(\cdot)|^p / \delta^\alpha \, dx \right)^{1/p}$$

it follows that the constant in (5.1.11) is sharp. $\qquad\qquad\square$

Remark 5.3

In [8], Theorem 4, it is proved that if $\mathbb{R}^n \setminus \Omega$ is a compact set, then for any $p \in [1, \infty)$ and $s \in [n, \infty)$,

$$\mu(p, s, \Omega) := \sup_{f \in C_0^\infty(\Omega), f \neq 0} \frac{\int_\Omega \frac{|f(x)|^p}{\delta(x)^s} \, dx}{\int_\Omega \frac{|\nabla f(x)|^p}{\delta(x)^{s-p}} \, dx} = \frac{|s - n|^p}{p^p}.$$

Hence with $s = p \geq n$,

$$\mu(p, \Omega) := \sup_{f \in C_0^\infty(\Omega), f \neq 0} \frac{\int_\Omega \frac{|f(x)|^p}{\delta(x)^p} \, dx}{\int_\Omega |\nabla f(x)|^p \, dx} = \left(\frac{p - n}{p} \right)^p.$$

The constant $(p - n)^p / p^p$ is therefore sharp in (5.1.12) whenever $\mathbb{R}^n \setminus \Omega$ is compact.

The assertion is false when $\mathbb{R}^n \setminus \Omega$ is unbounded. The case of a half-space Ω provides a counterexample for then $\mu(p, s, \Omega) = |s - 1|^p / p^p$.

5.2 Geometric Properties of Ω

The existence of an inequality (5.1.10) for some positive constant $C(p, \Omega)$ depends on the geometry of Ω and the nature of its boundary. In [72] and [60] a detailed study is made of the class of so-called *Generalised Ridged Domains*, this being a wide class which includes ones with special features, like horns, spirals and domains with fractal boundaries. The study includes an analysis of subsets of Ω which are significant for our present purposes; these are the so called *skeleton* and *ridge*. We refer to [15], Chapter 2 for background information and a detailed discussion of the results relevant to our needs in this chapter.

Let $N(x) := \{y \notin \Omega : |x - y| = \delta(x)\}$, and call it the *near set* of x on $\Omega^c := \mathbb{R}^n \setminus \Omega$. The *skeleton* of Ω is defined to be the subset

$$S(\Omega) := \{x \in \Omega : \operatorname{card} N(x) > 1\}, \tag{5.2.1}$$

where $\operatorname{card} N(x)$ denotes the cardinality of $N(x)$. Thus if $x \notin S(\Omega)$, there exists a unique $y \in N(x)$ and $\delta(x) = |x - y|$. From [60], Theorem 5.1.5, the function δ is differentiable at x if and only if $x \notin S(\Omega)$ and

$$\nabla \delta(x) = (x - y)/|x - y|, \quad x \in \Omega \setminus S(\Omega); \tag{5.2.2}$$

furthermore, $\nabla \delta$ is continuous on its domain of definition. Therefore, $S(\Omega)$ is the set of points in Ω at which δ is not differentiable. The function δ is uniformly Lipschitz on Ω; for let $x, y \in \Omega$ and choose $z \in \partial\Omega$ such that $\delta(y) = |y - z|$. Then

$$\delta(x) \leq |x - z| \leq |x - y| + \delta(y),$$

which together with the inequality obtained by reversing x and y yields

$$|\delta(x) - \delta(y)| \leq |x - y|.$$

Since a Lipschitz function is differentiable almost everywhere by Rademacher's theorem, it follows that $S(\Omega)$ is of zero Lebesgue measure. Also by (5.2.2) $|\delta(x)| = 1$ a.e. on Ω.

For $x \in \Omega$ and $y \in N(x)$, let

$$\lambda := \sup\{t \in (0, \infty) : y \in N(y + t[x - y])\}. \tag{5.2.3}$$

Then, for all $t \in (0, \lambda)$, $N(y + t[x - y]) = y$. The point $p(x) = y + \lambda(x - y)$ is called the *ridge point* of x in Ω and the *ridge* of Ω is defined by

$$R(\Omega) := \{p(x) : x \in \Omega\}. \tag{5.2.4}$$

Another important subset of Ω relevant to us is $\Sigma(\Omega) := \Omega \setminus G(\Omega)$, where $G(\Omega)$ is the *good set* defined by Li and Nirenberg in [129] as the largest open subset of Ω such that every point $x \in G(\Omega)$ has a unique near point.

The following connections between the sets $S(\Omega), R(\Omega), \Sigma(\Omega)$ are established in [15], Lemma 2.2.8:

$$S(\Omega) \subseteq R(\Omega) \subseteqq \overline{S(\Omega)},$$

$$\Sigma(\Omega) = \overline{R(\Omega)} = \overline{S(\Omega)}. \qquad (5.2.5)$$

In [85] Fremlin shows that $R(\Omega)$ coincides with the *central set* $R_C(\Omega)$ of centres of maximal open balls contained in Ω. It is also proved in [85] that for any proper open subset Ω of $\mathbb{R}^2$, $R(\Omega)$ has zero two-dimensional Lebesgue measure, but it does not appear to be known if this is the case for general open subsets of $\mathbb{R}^n$ for $n > 2$. An example is given in [131], page 10, of a convex open subset Ω of $\mathbb{R}^2$ with a $C^{1,1}$ boundary which is such that $\overline{S(\Omega)}$ has nonzero Lebesgue measure; hence $R(\Omega)$ is not closed in view of (5.2.4) and Fremlin's result. It is proved in [104] and [129] that $R(\Omega)$ is closed and is of zero measure if Ω is a domain in $\mathbb{R}^2$ with a $C^{2,1}$ boundary.

Bunt [39] and Motzkin [139] established independently the important result that

$$R(\Omega) = S(\Omega) = \varnothing, \qquad (5.2.6)$$

if and only if $\mathbb{R}^n \backslash \Omega$ is convex. The following statements are therefore equivalent (see [15], Theorem 2.2.9):

1. $\mathbb{R}^n \setminus \Omega$ is convex;
2. δ is differentiable at every $x \in \Omega$;
3. for every $x \in \Omega$, there is a unique point $y \in \mathbb{R}^n \setminus \Omega$ at minimal distance from x; thus $N(x) = \{y\}$.

In Section 1.3.1, for an open subset Ω of $\mathbb{R}^n (n \geq 2)$ with non-empty boundary $\partial \Omega$, the smoothness class $C^{k,\alpha}, k \in \mathbb{N}_0, \alpha \in [0, 1]$, of the boundary was defined. The smoothness of the boundary of Ω is reflected in that of the distance function δ. For instance, if $\partial \Omega \in C^k = C^{k,0}, k \geq 2$, then for some positive constant $\mu, \delta \in C^k(\Gamma_\mu)$, where $\Gamma_\mu = \{x \in \overline{\Omega} : \delta(x) < \mu\}$; see [89], Lemma 1 in the Appendix. Hence for every $x \in \Gamma_\mu$, there is a unique near point $y \in N(x)$ and consequently $\Gamma_\mu \subset G(\Omega)$, the good set. The same applies for the boundary smoothness condition $\partial \Omega \in C^{k,\alpha}, k \geq 1, 0 \leq \alpha \leq 1$. We refer to [89] for a full discussion of smoothness conditions on Ω and its boundary.

Let Ω be a domain in $\mathbb{R}^n (n \geq 2)$ with a C^2 boundary, thus locally, after a rotation of co-ordinates, $\partial \Omega$ is the graph of a C^2 function. To be specific, for any $y \in \partial \Omega$, let $\mathbf{n}(y), T(y)$ denote respectively the unit inward normal to $\partial \Omega$ at y and the tangent plane to Ω at y. The $\partial \Omega$ is of class C^2 if, given any $y_0 \in \partial \Omega$, there exists a neighbourhood $\mathcal{N}(y_0)$ in which $\partial \Omega$ is given in terms of local co-ordinates by $x_n = \phi(x_1, ..., x_{n-1}), \phi \in C^2 (T(y_0) \cap \mathcal{N}(y_0))$, where x_n lies in the direction of $\mathbf{n}(y_0)$ and with $x' = (x_1, ..., x_{n-1})$, we have

$$\mathbf{D}\phi(y_0') = (D_1, D_2, ..., D_{n-1})\,\phi(y_0')$$
$$= \left[\left(\partial/\partial x_1, \partial/\partial x_2, ..., \partial/\partial x_{(n-1)}\right)\phi\right](y_0') = 0.$$

The *principal curvatures* $\kappa_1, ..., \kappa_{n-1}$, of $\partial\Omega$ at y_0 are the eigenvalues of the Hessian matrix

$$\left[\mathbf{D}^2\phi(y_0')\right] = \left(D_i D_j \phi(y_0')\right)_{i,j=1,...,n-1}.$$

For a domain Ω in $\mathbb{R}^n$, $n \geq 2$, with C^2 boundary, it is proved in [15], Lemma 2.4.2 that $\delta \in C^2(G(\Omega))$, where by (5.2.5), $G(\Omega) = \Omega \setminus \overline{R(\Omega)} = \Omega \setminus \overline{S(\Omega)}$, and

$$\Delta\delta(x) = \Sigma_{i=1}^{n-1}\left(\frac{\kappa_i(y)}{1 + \delta(x)\kappa_i(y)}\right), \quad x \in G(\Omega), \ N(x) = \{y\};$$

here $\kappa_i(y)$, $i = 1, 2, ..., n-1$ are the principal curvatures of $\partial\Omega$ at y with respect to the unit inward normal. Moreover, with *mean curvature* defined by

$$H(y) := \frac{1}{n-1}\Sigma_{j=1}^{n-1}\kappa_j(y), \quad y \in \partial\Omega,$$

it is proved in [15], Propositions 2.5.3 and 2.5.4 that for $x \in G(\Omega)$ and $N(x) = y$, we have

$$1 + \delta(x)H(y) > 0$$

and

$$\Delta\delta(x) \leq \frac{(n-1)H(y)}{1 + \delta(x)H(y)}.$$

5.3 Convex Domains

Proofs of the following two important properties of δ for a convex domain Ω may be found in [15], Section 2.3:

1. δ is concave, i.e., for any $x, y \in \Omega$ and $z = \lambda x + (1 - \lambda)y$, where $\lambda \in (0, 1)$,

$$\delta(z) \geq \lambda\delta(x) + (1 - \lambda)\delta(y);$$

2. δ is superharmonic, i.e., $-\Delta\delta \geq 0$ in the distributional sense,

$$-\int_\Omega \delta(x)\Delta\phi(x)\,dx = -\int_\Omega \Delta\delta(x)\phi(x)\,dx \geq 0, \ \left(0 \leq \phi \in C_0^\infty(\Omega)\right).$$
$$(5.3.1)$$

For a domain Ω in $\mathbb{R}^n(n \geq 2)$ with a C^2 boundary, it is proved in [124] (see also [15], Proposition 2.5.4) that δ is superharmonic in the good subset $G(\Omega)$ of Ω if and only if Ω is weakly mean convex.

Since $|\delta(x)| = 1$ a.e. on Ω, we have for all non-negative $\phi \in C_0^\infty(\Omega)$ and $1 < p < \infty$,

$$\int_\Omega |\nabla\delta|^{p-2}\nabla\delta \cdot \nabla\phi\,dx = \int_\Omega \nabla\delta \cdot \nabla\phi\,dx = -\int_\Omega \delta\Delta\phi\,dx \geq 0, \qquad (5.3.2)$$

and hence if δ is superharmonic, the p-Laplacian satisfies

$$-\Delta_p\delta = -\text{div}\left(|\nabla\delta|^{p-2}\nabla\delta\right) \geq 0 \qquad (5.3.3)$$

in the distributional sense; δ is then said to be *p-superharmonic* on Ω. It follows that

$$\int_\Omega |\nabla\delta|^{p-2}\nabla\delta \cdot \nabla\phi\,dx = \int_\Omega \nabla\delta \cdot \nabla\phi\,dx$$

for $0 \leq \phi \in C_0^1(\Omega)$.

The fact that δ is p-superharmonic on a convex domain Ω implies the validity of an inequality (5.1.10) on Ω with $C(p,\Omega) = (p/(p-1))^p$. To see this, we follow a trick of Moser in [138]. Let $\phi = |u|^p/\delta^{p-1}$ in (5.3.2), with $u \in C_0^\infty(G(\Omega))$, where $G(\Omega)$ is the *good set* in Ω. Since δ is differentiable on $G(\Omega)$, we have that $\phi \in C_0^1(G(\Omega))$ and

$$p\int_\Omega (\nabla\delta \cdot \nabla|u|) \frac{|u|^{p-1}}{\delta^{p-1}}\,dx - (p-1)\int_\Omega \frac{|u|^p}{\delta^p}|\nabla\delta|^2 dx \geq 0.$$

Hence, as $|\nabla|u|| \leq |\nabla u|$ a.e. and $|\nabla\delta| = 1$ on $G(\Omega)$, we have

$$(p-1)\int_\Omega \frac{|u|^p}{\delta^p}\,dx \leq p\int_\Omega |\nabla u|\frac{|u|^{p-1}}{\delta^{p-1}}\,dx$$

$$\leq p\left(\int_\Omega |\nabla u|^p dx\right)^{1/p}\left(\int_\Omega \frac{|u|^p}{\delta^p}\,dx\right)^{1-1/p},$$

and so

$$\int_\Omega \left|\frac{u(x)}{\delta(x)}\right|^p dx \leq \left(\frac{p}{p-1}\right)^p \int_\Omega |\nabla u(x)|^p\,dx, \quad u \in \overset{0}{D}_p^1(G(\Omega)). \qquad (5.3.4)$$

This is an example of a Hardy inequality which holds for functions defined on a subset $\Gamma_r := \{x \in \Omega: \delta(x) < r\}$ of Ω. In [155], Robinson investigates the general question of whether, for $s \geq 0$, $r_0 > 0$, $r \in (0, r_0)$ and $a > 0$, the weighted $L_2(\Omega)$ Hardy inequality

$$\int_\Omega \delta(x)^{s-2}|f(x)|^2\,dx \leq a^2 \int_\Omega \delta(x)^s|\nabla f(x)|^2\,dx \qquad (5.3.5)$$

is valid for all $f \in C_0^1(\Gamma_r)$, where $\Gamma_r = \{x \in \Omega: \delta(x) < r\}$. The Hardy constant $a_s(\Gamma_r)$ is then defined to be the infimum of all the constants a for which (5.3.5) is satisfied. It clearly decreases as $r \to 0$ and, denoting the boundary of Ω by

Γ, the *boundary constant* $a_s(\Gamma)$ is defined as the infimum of $a_s(\Gamma_r)$ over $r \in (0, r_0)$. The inequalities (5.3.4) and (5.3.5) are *boundary* Hardy inequalities, in the sense that they do not hold for functions on all of Ω. The example $\Omega = B_1(0)$ demonstrates that this is all that can be achieved in general. For (5.3.5) is then valid for $s \in [0, 1)$ with $a_s(\Omega) = 2/(1 - s)$ whereas if $s > 1$, (5.3.5) holds on $C_0^1(\Gamma_r)$ for all $r \in (0, 1)$ but fails on $C_0^1(B_1(0))$; see [113].

In [155], Theorems 4.3 and 5.1, the precise value of the boundary Hardy constant $a_s(\Gamma)$ is determined under the assumption that Ω is either convex or a $C^{1,1}$ domain. The assumption that Ω is a $C^{1,1}$ domain implies that its boundary satisfies a uniform internal ball condition and a uniform external ball condition. The uniform internal ball condition requires that for each $y \in \partial\Omega$, there exists $x \in \Omega$ and $k > 0$ such that $\overline{B}(x; k) \cap \Omega^c = \{y\}$. Hence, for small enough r, $\Gamma_r \subset G(\Omega)$, the *good set* of Ω. The uniform exterior ball condition is similar with Ω and Ω^c interchanged.

Theorem 5.4 *Let Ω be either convex or a $C^{1,1}$ domain in $\mathbb{R}^n$. Then for all $r \in (0, r_0)$ with r_0 sufficiently small, and all s such that $0 \leq s \neq 1$,*

$$\int_\Omega \delta(x)^{s-2} |f(x)|^2 \, dx \leq a_s(\Gamma_r)^2 \int_\Omega \delta(x)^s |\nabla f(x)|^2 \, dx \qquad (5.3.6)$$

for all $f \in C_0^1(\Gamma_r)$. Moreover, the Hardy boundary constant is $a_s(\Gamma) = \frac{2}{|s-1|}$.

The boundary constant $a_s(\Gamma)$ is characterised by local constants in the sense that

$$a_s(\Gamma) = \sup_{j \in \mathbb{N}} a_s(\Gamma \cap U_j),$$

where $(U_j)_{j \in \mathbb{N}}$ is a cover of Γ by bounded open subsets of $\mathbb{R}^n$.

In [155] the inequality (5.3.6) is shown to be equivalent to a weighted version of Davies' *weak Hardy inequality* in [48], with equality of the corresponding optimal constants. The weak Hardy inequality on $C_0^1(\Omega)$ is

$$\int_\Omega \delta(x)^{s-2} |\psi(x)|^2 \, dx \leq b^2 \int_\Omega \delta(x)^s(x) |\nabla\psi(x)|^2 \, dx + c^2 \int_\Omega |\psi(x)|^2 \, dx \quad (5.3.7)$$

for all $\psi \in C_0^1(\Omega)$ and some finite constants b, c. The *weak* Hardy constant $b_s(\Omega)$ is defined to be the infimum of all the b for which there is a c such that (5.3.7) is valid. One can also define $b_s(\Gamma_r)$ and $b_s(\Gamma)$ by restriction to functions in $C_0^1(\Gamma_r)$, as was done for $a_s(\Gamma_r)$. The aforementioned equivalence is given in

Theorem 5.5 *Let $s \in [0, 2)$. Then the boundary Hardy inequality (5.3.6) is valid if and only if the weak Hardy inequality (5.3.7) on $C_0^1(\Gamma_r)$ is valid. Moreover, if the inequalities are valid then $a_s(\Gamma) = b_s(\Gamma) = b_s(\Omega)$.*

For a convex domain Ω with a C^1 boundary, (5.3.4) was proved in [133] to hold for all $u \in C_0^\infty(\Omega)$ and

$$l(p, \Omega) := \inf_{u \in W_0^{1,p}(\Omega)} \frac{\int_\Omega |\nabla u|^p \, dx}{\int_\Omega |u/\delta|^p \, dx} = \left(\frac{p-1}{p}\right)^p =: c_p. \tag{5.3.8}$$

This was also established in [132] assuming only that Ω is convex. The existence of a minimiser in the variational problem determined by (5.3.8) was also explored in [132] for a bounded Ω with a C^2 boundary. For $1 < p < \infty$, it was shown that $l(p, \Omega) \le \left(\frac{p-1}{p}\right)^p$, with equality if there is no minimiser; if $p = 2$, there is equality if and only if there is no minimiser. The existence of minimisers of (5.3.8) for domains of class $C^{1,\gamma}$, $\gamma \in [0, 1]$ is an important feature of [113]. It is proved in [113], Theorem 4.1, that if Ω is bounded and $l(p, \Omega) < c_p$, then there exists a positive minimiser $u \in \overset{0}{W}_{1,p}(\Omega)$ of (5.3.8). Also, if $\alpha \in ((p-1)/p, 1)$ is such that $l(p, \Omega) = \lambda_\alpha$, then $0 < u(x) < C\delta(x)^\alpha$ for all $x \in \Omega$. The identity (5.3.8) is proved in [124] for a domain Ω which is weakly mean convex in $\mathbb{R}^n (n \ge 2)$, as long as the set $\Sigma(\Omega) = \Omega \setminus G(\Omega)$ is assumed to have zero measure. The weak mean convexity condition is sharp in the sense that the equality fails if only the mean curvature $H \le \varepsilon$ is assumed for $\varepsilon > 0$.

In [154], Proposition 2.5, the following L_p version of (5.3.6) is given. Let $s \ge 0$ and suppose that $p - 1 - s > 0$. Then

$$\int_\Omega \delta(x)^{s-p} |f(x)|^p \, dx \le a_p^p \int_\Omega \delta(x)^s |\nabla f(x)|^p \, dx, \quad f \in C_0^1(G(\Omega)), \tag{5.3.9}$$

where $a_p = (p/[p-1-s])$.

5.3.1 Convex Complements

The conclusion of Theorem 5.4 continues to be correct if Ω is the complement of a convex set and $s > 1$, but if $s \in [0, 1)$ the constant $a_s(\Omega)$ can be strictly larger than $2/|s-1|$. The following analogue of (5.3.6) is derived in [154] for $\Omega = \mathbb{R}^n \setminus K$, where K is a closed convex subset of $\mathbb{R}^n$. The existence of the inequality is given by Theorem 1.1 in [154] and the optimality of the derived constant in Theorem 4.2.

Theorem 5.6 Let $\Omega = \mathbb{R}^n \setminus K$ ($n \ge 2$), where K is a closed convex subset of $\mathbb{R}^n$, and denote the Hausdorff dimension of the boundary $\partial\Omega$ of Ω by d_H. Let $c_\Omega = c \circ d_H$, where $c(t) = t^s(1+t)^{s-s'}$ with $s, s' \ge 0$. If $n - d_H + (s \wedge s') - p > 0$, with $p \in [1, \infty)$ and $s \wedge s' := \max\{s, s'\}$, then for all $\phi \in C_0^1(\Omega)$,

$$\int_\Omega c_\Omega |\nabla\phi(x)|^p dx \ge \int_\Omega c_\Omega |(\nabla\delta(x)) \cdot (\nabla\phi(x))|^p \, dx \ge a_p^p \int_\Omega c_\Omega \frac{|\phi(x)|^p}{\delta(x)^p} dx, \tag{5.3.10}$$

where $a_p = (n - d_H + (s \wedge s') - p)^p$.

Let $d_H \in \{1, ..., n-1\}$ and define the optimal constant

$$l(p, \Omega) = \inf \left\{ \frac{\int_\Omega c_\Omega |\nabla \phi(x)|^p dx}{\int_\Omega c_\Omega \frac{|\phi(x)|^p}{\delta(x)^p} dx} : \phi \in C_0^\infty(\Omega) \right\}.$$

Then

$$l(p, \Omega) \leq ([n - d_H + s - p]/p)^p,$$

with equality if $s \leq s'$.

Another weighted inequality on a domain with a convex complement is the following from [8], Theorem 3.

Theorem 5.7 *Let $\Omega = \mathbb{R}^n \backslash K$ ($n \geq 2$), where K is a closed, non-empty, convex subset of $\mathbb{R}^n$. Then for any $p \in [1, \infty)$, $s \in \mathbb{R}$ and real-valued $u \in C_0^1(\Omega)$,*

$$\int_\Omega \frac{|\nabla u(x)|^p}{\delta^{s-p}(x)} dx \geq c_{n,p,s} \int_\Omega \frac{|u(x)|^p}{\delta^s(x)} dx, \qquad (5.3.11)$$

where

$$c_{n,p,s} = \min\{|s - k|^p/p^p : k = 1, 2, ..., n\} \qquad (5.3.12)$$

is optimal. Hence with $s = p \in [n, \infty)$,

$$\inf_{u \in C_0^1(\Omega), u \neq 0} \frac{\int_\Omega |\nabla u(x)|^p dx}{\int_\Omega \frac{|u(x)|^p}{\delta^p(x)} dx} \geq (p - n)^p/p^p. \qquad (5.3.13)$$

Note from Remark 5.3 that by Theorem 4 in [8], if K in Theorem 5.7 is assumed to be a non-empty compact subset of $\mathbb{R}^n$ rather than convex, then for any $p \in [1, \infty)$ and $s \in [n, \infty)$,

$$c_{n,p,s} = |s - n|^p/p^p. \qquad (5.3.14)$$

5.3.2 Non-convex Domains

For any domain Ω in $\mathbb{R}^n (n \geq 1)$, it is proved in [12] that Hardy's inequality

$$\int_\Omega \frac{|f(x)|^2}{\delta(x)^2} dx \leq C(\Omega) \int_\Omega |\nabla f(x)|^2 dx, \quad f \in C_0^\infty(\Omega) \qquad (5.3.15)$$

holds for a finite constant $C(\Omega)$ if and only if there exist a strictly positive superharmonic function g on Ω and a positive number ε such that

$$\Delta g + \frac{\varepsilon}{\delta^2} g \leq 0 \qquad (5.3.16)$$

in the distributional sense, i.e.,

$$\int_\Omega \left(\Delta g + (\varepsilon/\delta^2) g \right) \psi \, dx = \int_\omega \left(\Delta \psi + (\varepsilon/\delta^2) \psi \right) g \, dx \leq 0, \quad 0 \leq \psi \in C_0^\infty(\Omega).$$

The largest value of ε in (5.3.16) is $1/C(\Omega)$, where $C(\Omega)$ is the best possible constant in (5.3.15). The function g is a so-called 'strong barrier' on Ω. We refer to [12] for background information and the proof of this important result.

For non-convex domains (and ones not weakly mean convex), the best possible constant in the Hardy inequality is not known in general, but for arbitrary planar, simply connected domains Ω, there is the following celebrated result of Ancona in [12]:

Theorem 5.8 *Let $\Omega \subsetneq \mathbb{R}^2$ be a simply connected domain. Then (5.3.15) holds with $C(\Omega) \leq 16$.*

Since the optimal constant $C(\Omega)$ in (5.3.15) (which, for now, we call the *strong* Hardy constant to distinguish it from the *weak* Hardy constant of Section 5.3) is 4 for a convex planar domain Ω, it is natural to ask if $C(\Omega)$ can take values between 4 and 16 if Ω possesses some degree of convexity. This was answered by Laptev and Sobolev in [115]; they established a refinement of Kobe's '1/4' theorem and introduced two possible 'measures' of non-convexity in their solution. In particular they proved that if any $y \in \partial\Omega$ is the vertex of an infinite sector Λ of angle $\theta \in [\pi, 2\pi]$ independent of y such that $\Omega \subset \Lambda$, then $C(\Omega)$ in Theorem 5.8 can be replaced by $4\theta^2/\pi^2$. The convexity case corresponds to $\theta = \pi$ and then 4 is recovered for $C(\Omega)$.

In [48] Davies determined the value of $C(\Omega_\beta)$ for the plane sector

$$\Omega_\beta := \{re^{i\theta} : 0 < r < 1, \ 0 < \theta < \beta\}, \ \ 0 < \beta < 2\pi.$$

He proved that the strong and weak Hardy constants are equal whenever $0 < \beta < 2\pi$. Furthermore, denoting the common value by C_β, there exists a critical angle $\beta^c = 4.856$ such that $C_\beta = 4$ for all $\beta \leq \beta^c$, while for $\beta^c < \beta \leq 2\pi$, C_β is strictly increasing and $4 < C_\beta \leq C_{2\pi} = 4.869$.

The example of a quadrilateral Ω in $\mathbb{R}^2$ with exactly one non-convex angle β, $\pi < \beta < 2\pi$ is considered in [18]. The best possible constant C_β is shown to be the unique solution of

$$2\sqrt{C_\beta} \left(\frac{\Gamma\left(\frac{3+\sqrt{1-4C_\beta}}{4}\right)}{\Gamma\left(\frac{1+\sqrt{1-4C_\beta}}{4}\right)} \right)^2 = \tan\left(\sqrt{C_\beta}\left(\frac{\beta - \pi}{2}\right)\right)$$

when $\beta^c \leq \beta < 2\pi$ and $C_\beta = 1/4$ when $\pi < \beta \leq \beta^c$. The constant C_β is precisely that computed numerically in [48] for a sector of angle β. The critical angle β^c is the unique solution in $(\pi, 2\pi)$ of the equation

$$\tan\left(\frac{\beta^c - \pi}{4}\right) = 4\left(\frac{\Gamma(\frac{3}{4})}{\Gamma(\frac{1}{4})}\right)^2.$$

The Hardy constant for other non-convex planar domains is computed in [19].

Hardy inequalities on annular regions bounded by convex domains with smooth boundaries were investigated by Avkhadiev and Laptev in [5]. An analogue of Theorem 1 in [5] is used in [15], Section 3.8 to derive a Hardy inequality on a general doubly connected domain $\Omega \subset \mathbb{R}^2 \equiv \mathbb{C}$. Such a domain has a boundary which is the disjoint union of 2 simple curves. If its boundary is smooth then it can be mapped conformally onto an annulus $\Omega_{\rho,R} = B_R \setminus B_\rho = \{z \in \mathbb{C} : \rho < |z| < R\}$ for some ρ, R. Let $\Omega := \Omega_2 \setminus \overline{\Omega_1} \subset \mathbb{C}$ and $B_\rho \subset B_R \subset \mathbb{C}, 0 < \rho < R$, where B_r is the disc of radius r centred at the origin. Let

$$F \colon \Omega_2 \setminus \overline{\Omega_1} \to B_R \setminus \overline{B_\rho}$$

be analytic and univalent. Then in [15], Lemma 3.8.3, it is shown that

$$\mathcal{F}(z) := -\frac{|F'(z)|^2}{|F(z)|^2} + |F'(z)|^2 \left\{ \frac{1}{|F(z)| - \rho} + \frac{1}{R - |F(z)|} \right\}^2$$

is invariant under scaling, rotation and inversion, which implies it does not depend on the mapping F but only on the geometry of $\Omega_2 \setminus \overline{\Omega_1}$. Theorem 1 in [5] can then be shown to yield

Theorem 5.9 *For* $\Omega := \Omega_2 \setminus \overline{\Omega_1} \subset \mathbb{R}^2$,

$$\int_\Omega |\nabla u(x)|^2 dx \geq \frac{1}{4} \int_\omega \mathcal{F}(x) |u(x)|^2 dx, \quad u \in \overset{0}{H}_2^1(\Omega). \tag{5.3.17}$$

Avkhadiev proves in [4] that for $\Omega := B_R \setminus \overline{B_\rho}$,

$$\int_\Omega |\nabla u(x)|^2 dx \geq \lambda(\Omega) \int_\Omega \frac{|u(x)|^2}{\delta(x)^2} dx, \quad u \in \overset{0}{H}_2^1(\Omega), \tag{5.3.18}$$

where

$$\frac{2}{\pi} \ln \frac{R}{\rho} \leq \frac{1}{\lambda(\Omega)} \leq \ln \frac{R}{\rho} + k_0,$$

and $k_0 = \Gamma(\frac{1}{4})^4 / 2\pi^2 = 8.75\dots$ This inequality is applied in [15], Example 3.8.7, to prove that for $n \geq 3$ and any $\varepsilon > 0$, there exist ellipsoids E_1, E_2 with $\overline{E_2} \subset E_1 \subset \mathbb{R}^n$ and a function $f \in C_0^1(E_1 \setminus \overline{E_2})$ such that

$$\int_{E_1 \setminus \overline{E_2}} |\nabla f(x)|^2 dx \leq \varepsilon \int_{E_1 \setminus \overline{E_2}} \frac{|f(x)|^2}{\delta(x)^2} dx,$$

where $\delta(x)$ is the distance from $x \in E_1 \setminus E_2$ to the boundary of $E_1 \setminus E_2$. Moreover, the mean curvature $H(N(x)) \leq \varepsilon$ for all $x \in E_1 \setminus \overline{E_2}$, $N(x)$ being the near point of x.

5.4 The Mean Distance Function

Let Ω be a domain in $\mathbb{R}^n$, $n \geq 2$ with non-empty boundary, and for $x \in \Omega$, $v \in \mathbb{S}^{n-1}$, set

$$\tau_v(x) = \min\{t > 0 : x + tv \notin \Omega\}, \quad \delta_v(x) = \min\{\tau_v(x), \tau_{-v}(x)\}.$$

The *mean distance function* M_p is defined by

$$\frac{1}{M_p(x)^p} := \frac{\sqrt{\pi}\,\Gamma\left(\frac{n+p}{2}\right)}{\Gamma\left(\frac{p+1}{2}\right)\Gamma\left(\frac{n}{2}\right)} \int_{\mathbb{S}^{n-1}} \frac{1}{\delta_v^p(x)}\, d\omega(v), \tag{5.4.1}$$

where the measure $d\omega(v)$ on $\mathbb{S}^{n-1}$ is assumed to be normalised, i.e., $\int_{\mathbb{R}^n} d\omega(v) = 1$. It was introduced by Davies in [47] for $p = 2$ and for any $p \in (0, \infty)$ in [167]. For background information we refer to [15], 3.3. It is an effective and much used tool for establishing Hardy and similar inequalities in two and higher dimensions by reduction to one-dimensional problems. If Ω has suitable geometric properties the mean distance function can be estimated in a useful way. For example, when $\partial\Omega$ satisfies a θ-cone condition (every $x \in \partial\Omega$ is the vertex of a circular cone of semi-angle θ that lies entirely in $\mathbb{R}^n \backslash \Omega$), the mean distance function is equivalent to the usual distance function (see [15], p. 85); when Ω is convex, it is shown below that it is bounded above by the ordinary distance function. The following theorem from [47] (for $p = 2$) and [167] demonstrates its use.

Theorem 5.10 *For all $f \in \overset{0}{\mathcal{D}}_p^1(\Omega)$, $1 < p < \infty$ and any domain Ω with non-empty boundary,*

$$\int_\Omega \left|\frac{f(x)}{M_p(x)}\right|^p dx \leq \left(\frac{p}{p-1}\right)^p \int_\Omega |\nabla f(x)|^p\, dx, \quad f \in \overset{0}{\mathcal{D}}_p^1(\Omega). \tag{5.4.2}$$

Proof The starting point is the one-dimensional inequality

$$\int_a^b |\phi'(t)|^p dt \geq \left(\frac{p-1}{p}\right)^p \int_a^b \frac{|\phi(t)|^p}{\rho(t)^p}\, dt, \quad \phi \in C_0^\infty(a, b), \tag{5.4.3}$$

where $\rho(t) = \min\{|t - a|, |t - b|\}$. Let ϕ be real and $c := (a + b)/2$. Then

$$\begin{aligned}
\int_a^c \frac{|\phi(t)|^p}{(t-a)^p}\, dt &= \int_a^c (t-a)^{-p} \left(\int_a^t [|\phi(x)|^p]' dx\right) dt \\
&= \int_a^c [|\phi(x)|^p]' \left(\int_x^c (t-a)^{-p} dt\right) dx \\
&\leq \frac{p}{p-1} \int_a^c \frac{|\phi(x)|^{p-1}|\phi'(x)|}{(x-a)^{p-1}}\, dx
\end{aligned}$$

since $||\phi(x)|'| \le |\phi'(x)|$ a.e. Similarly,

$$\int_c^b \frac{|\phi(t)|^p}{(t-a)^p} dt \le \frac{p}{p-1} \int_c^b \frac{|\phi(x)|^{p-1}|\phi'(x)|}{(b-x)^{p-1}} dx.$$

The two inequalities combine to give

$$\int_a^b \frac{|\phi(t)|^p}{\rho(t)^p} dt \le \left(\frac{p}{p-1}\right) \int_a^b \frac{|\phi(x)|^{p-1}|\phi'(x)|}{(\rho(x))^{p-1}} dx$$

$$\le \left(\int_a^b \frac{|\phi(t)|^p}{\rho(t)^p} dt\right)^{1-1/p} \left(\int_a^b |\phi'(x)|^p dx\right)^{1/p}$$

whence (5.4.3).

For $v \in \mathbb{S}^{n-1}$ and $\partial_v := v \cdot \nabla$, let (a_v, b_v) be the interval of intersection of Ω with the ray in direction v, $\delta_v(t) := \min\{|t - a_v|, |b_v - t|\}$, and denote by $\langle v, \omega \rangle$ the angle between v and $\omega \in \mathbb{R}^n$. Then from (5.4.3),

$$\int_{a_v}^{b_v} |\partial_v \phi(t)|^p dt \ge \left(\frac{p-1}{p}\right)^p \int_{a_v}^{b_v} \frac{|\phi(t)|^p}{\delta_v(t)^p} dt, \quad \phi \in C_0^\infty(a_v, b_v). \quad (5.4.4)$$

On integrating both sides with respect to the normalised measure $d\omega(v)$ and writing $v \cdot \nabla\phi = |\nabla\phi| \cos\langle v, \nabla\phi \rangle$, we obtain

$$\int_\Omega \int_{\mathbb{S}^{n-1}} |\cos(v, \nabla\phi(x))|^p d\omega(v) |\nabla\phi(x)|^p dx$$

$$\ge \left(\frac{p-1}{p}\right)^p \int_\Omega \int_{\mathbb{S}^{n-1}} \frac{1}{\delta_v(x)^p} d\omega(v) |\phi(x)|^p dx. \quad (5.4.5)$$

For any fixed unit vector $\mathbf{e} \in \mathbb{R}^n$,

$$\int_{\mathbb{S}^{n-1}} |\cos\langle v, \nabla\phi(x) \rangle|^p d\omega(v) = \int_{\mathbb{S}^{n-1}} |\cos\langle v, \mathbf{e} \rangle|^p d\omega(v)$$

and a calculation gives

$$\int_{\mathbb{S}^{n-1}} |\cos\langle v, \mathbf{e} \rangle|^p d\omega(v) = \frac{\Gamma\left(\frac{p+1}{2}\right) \Gamma\left(\frac{n}{2}\right)}{\sqrt{\pi} \Gamma\left(\frac{n+p}{2}\right)}. \quad (5.4.6)$$

The inequality (5.4.2) follows from (5.4.5) for any real $\phi \in C_0^\infty(\Omega)$, and hence any real $\phi \in \mathcal{D}_p^1$. If ϕ is not real, then $|\phi| \in \mathcal{D}_p^1$, $|\nabla|\phi|| \le |\nabla\phi|$ a.e., and (5.4.2) is a consequence of the inequality already established for real functions. $\square$

Remark 5.11

In a case like $\Omega = \mathbb{R}^n \setminus \{0\}$, where $\rho_v(t) = \infty$ unless the ray v passes through the origin, a co-ordinate change is necessary; see the proof of Theorem 2.3 in [143]. Let $\{u_1, u_2, ..., u_n\}$, $u_1 = v$ be an orthonormal basis of $\mathbb{R}^n$, let

$\mathbf{v} = (v_1, v_2, ..., v_n)$ denote co-ordinates with respect to that basis and let P be a co-ordinate transition matrix $x = \mathbf{v}P$ from $\mathbf{v}$ co-ordinates to standard co-ordinates. For fixed $\hat{\mathbf{v}} = (v_2, ..., v_n)$ let $\Omega_{\hat{\mathbf{v}}} = \{v_1 \in \mathbb{R}: \mathbf{v}P \in \Omega\}$, where $\mathbf{v} = (v_1, \hat{\mathbf{v}})$ and $\Omega_{v_1} = \{\hat{\mathbf{v}}: \mathbf{v}P \in \Omega\}$. Define $g_{\hat{\mathbf{v}}}: \Omega_{\hat{\mathbf{v}}} \to \mathbb{R}$ and $\delta_v \hat{\mathbf{v}} \to (0, \infty]$ by

$$g_{\hat{\mathbf{v}}}(v_1) := f(\mathbf{v}P), \quad \delta_{\hat{\mathbf{v}}}(v_1) = \delta_{v_1}(\hat{\mathbf{v}}P).$$

Then from (5.4.4),

$$\int_{\Omega_{\hat{\mathbf{v}}}} |g'_{\hat{\mathbf{v}}}(v_1)|^p dv_1 \geq \left(\frac{p-1}{p}\right)^p \int_{\Omega_{\hat{\mathbf{v}}}} \frac{|g_{\hat{\mathbf{v}}}(v_1)|^p}{\delta_{\hat{\mathbf{v}}}(v_1)^p} dv_1$$

and hence

$$\left(\frac{p-1}{p}\right)^p \int_{\Omega} \frac{|f(x)|^p}{M_p(x)^p} dx = \left(\frac{p-1}{p}\right)^p \int_{\mathbb{S}^{n-1}} \int_{\Omega_{v_1}} \int_{\Omega_{\hat{\mathbf{v}}}} \frac{|g_{\hat{\mathbf{v}}}(v_1)|^p}{\delta_{\hat{\mathbf{v}}}(v_1)^p} dv_1 \, d\hat{\mathbf{v}} \, dv$$

$$\leq \int_{\mathbb{S}^{n-1}} \int_{\Omega_{v_1}} \int_{\Omega_{\hat{\mathbf{v}}}} |g'_{\hat{\mathbf{v}}}(v_1)|^p dv_1 \, d\hat{\mathbf{v}} \, dv$$

$$= \int_{\mathbb{S}^{n-1}} \int_{\Omega} |(v \cdot \nabla)f(x)|^p \, d\omega(v) \, dx,$$

which corresponds to (5.4.3) and hence leads to (5.4.2).

Theorem 5.12 *If Ω is convex, then $M_p(x) \leq \delta(x)$ for all $x \in \Omega$ and hence*

$$\int_{\Omega} \left|\frac{f(x)}{\delta(x)}\right|^p dx \leq \left(\frac{p}{p-1}\right)^p \int_{\Omega} |\nabla f(x)|^p \, dx, \quad f \in \overset{0}{D}{}^1_p(\Omega). \tag{5.4.7}$$

Proof Let $\mathbf{e}$ be a unit vector in $\mathbb{R}^n$ which is such that $\rho_{\mathbf{e}}(x) = \delta(x)$. Then as Ω is assumed to be convex,

$$\delta_v(x) \cos\langle\mathbf{e}, v\rangle \leq \delta(x).$$

Hence

$$\int_{\mathbb{S}^{n-1}} \frac{1}{\delta_v(x)^p} \, d\omega(v) \geq \int_{\mathbb{S}^{n-1}} |\cos\langle\mathbf{e}, v\rangle|^p \frac{1}{\delta(x)^p} \, d\omega(v)$$

$$= \frac{\Gamma\left(\frac{p+1}{2}\right)\Gamma\left(\frac{n}{2}\right)}{\sqrt{\pi}\Gamma\left(\frac{n+p}{2}\right)} \frac{1}{\delta(x)^p}$$

by (5.4.6) and so $M_p(x) \leq \delta(x)$ which yields (5.4.7) from (5.4.6). $\qquad\square$

Let $r := \sup\{\delta(x): x \in \Omega\}$ and $\mu := \sup\{M_2(x): x \in \Omega\}$ denote respectively the *inradius* and *mean inradius* of Ω. From (5.4.7) with $p = 2$, we have that the least eigenvalue λ_{Ω} of the Dirichlet Laplacian $-\Delta^D_{\Omega}$ on Ω satisfies

$$\lambda_\Omega = \inf\{\int_\Omega |\nabla u(x)|^2 \, dx : u \in \overset{0}{H}{}^1_2(\Omega), \int_\Omega |u(x)|^2 \, dx = 1\} \geq \frac{1}{4\mu^2}.$$

A lower bound for λ_Ω can also be obtained in terms of r. For if $\rho < r$ then Ω contains a ball B_ρ of radius ρ and

$$\lambda_\Omega \leq \inf\{\int_{B_\rho} |\nabla u(x)|^2 dx : u \in \overset{0}{H}{}^1_2(B_\rho), \int_{B_\rho} |u(x)|^2 dx = 1\}$$

$$= \left(1/\rho^2\right) \inf\{\int_{B_1} |\nabla u(x)|^2 dx : u \in \overset{0}{H}{}^1_2(B_1), \int_{B_1} |u(x)|^2 dx = 1\}$$

$$= \left(1/\rho^2\right) \lambda_1,$$

where $\lambda_1 := \lambda_{B_1}$ denotes the smallest eigenvalue of $-\Delta^D_{B_1}$. Therefore

$$\frac{1}{4\mu^2} \leq \lambda_\Omega \leq \frac{\lambda_1}{r^2}. \tag{5.4.8}$$

For $n \geq 2$, the value of λ_Ω is unchanged when Ω is punctured by a finite number of points (see [64], Corollary VIII.6.4), which means that μ is unaffected while r is reduced. Thus the mean inradius μ is of greater significance than the inradius r in the determination of λ_Ω for $n \geq 2$.

If Ω is convex, then $\mu \leq r$ by Theorem 5.12 and hence $\lambda_\Omega \geq 1/4r^2$. If Ω is mean convex, it is known that $\lambda_\Omega \geq c/\rho^2$, for some constant c, where ρ is the radius of the largest ball contained in Ω. This is not true for a general Ω but in [125] Lieb proved it to be true if the largest ball B_ρ contained in Ω is replaced by a ball that *intersects* Ω *significantly*. We reproduce an alternative proof of Lieb's result from [80] which uses the Hardy inequality (5.4.7) in the case $p = 2$ and the inequality, for any $x \in \Omega$ and $\rho > 0$,

$$|\Omega \cap B_\rho(x)| \leq \left(1 - \frac{\rho^2}{nM_2(x)^2}\right) |B_\rho(x)|, \tag{5.4.9}$$

where $B_\rho(x) = \{y : |y - x| < \rho, \}$, and

$$\frac{1}{M_2(x)^2} := \frac{n}{\omega_{n-1}} \left(\int_{\mathbb{S}^{n-1}} \frac{1}{\delta_\nu(x)^2} \, d\omega(\nu)\right), \quad \delta_\nu(x) := \inf\{|t| : x + t\nu \notin \Omega\}; \tag{5.4.10}$$

note the inclusion of ω_{n-1} to normalise the surface measure, as required in our definition of the mean distance function M_2. The estimate (5.4.9) is proved as follows. Denoting the characteristic function of Ω by χ_Ω, we have

$$|\Omega \cap B_\rho(x)| = \int_{\mathbb{S}^{n-1}} \int_0^\rho \chi_\Omega(x + t\nu) t^{n-1} dt \, d\omega(\nu).$$

For any $\nu \in \mathbb{S}^{n-1}$ with $\delta_\nu(x) > \rho$, we have $x + t\nu \in \Omega$ for all $t \in (0, \rho)$ and thus

$$|\Omega \cap B_\rho(x)| \geq |\{\nu \in \mathbb{S}^{n-1} : \delta_\nu(x) \geq \rho\}| n^{-1} \rho^n. \tag{5.4.11}$$

On the other hand,

$$\rho^{-2}|\{v \in \mathbb{S}^{n-1} : \delta_v(x) \le \rho\}| \le \int_{\mathbb{S}^{n-1}} \delta_v^{-2} d\omega(v) = \frac{\omega_{n-1}}{nM_2(x)^2},$$

and so

$$|\{v \in \mathbb{S}^{n-1} : \delta_v(x) \ge \rho\}| \ge \left(1 - \frac{\rho^2}{nM_2(x)^2}\right)\omega_{n-1}.$$

On inserting this in (5.4.11), (5.4.9) follows.

From (5.4.2),

$$\lambda_\Omega \ge \frac{1}{4}\inf\left\{\int_\Omega \frac{|u(x)|^2}{M_2(x)^2}dx : u \in \overset{0}{H}{}^1_2(\Omega), \int_\Omega |u(x)|^2 dx = 1\right\}$$

$$\ge \frac{1}{4}\inf\{M_2(x)^{-2} : x \in \Omega\}.$$

The lower bound for λ_Ω in [80] is given by using (5.4.9) in the last estimate, to give

Theorem 5.13 *Let $\Omega \subset \mathbb{R}^n$ be open. Then for any $\rho > 0$,*

$$\lambda_\Omega \ge \frac{1}{4\rho^2}\left(1 - \sup_{x\in\Omega}\frac{|\Omega \cap B_\rho(x)|}{|B_\rho(x)|}\right). \qquad (5.4.12)$$

This implies for all $\theta \in (0, 1)$,

$$\lambda_\Omega \ge \frac{(1-\theta)}{4\rho_\theta^2}, \; where \; \rho_\theta := \inf\left\{\rho > 0 : \sup_{x\in\Omega}\frac{|\Omega \cap B_\rho(x)|}{|B_\rho(x)|} \le \theta\right\}. \quad (5.4.13)$$

As remarked in [80], Theorem 5.13 has a counterpart for the principal eigenvalue of the p-Laplacian $-\Delta_{p,\Omega}$. From (5.4.1) the mean distance function is now given by

$$\frac{1}{M_p(x)^p} = C(n, p)\int_{\mathbb{S}^{n-1}} \frac{1}{\delta_v^p(x)}d\omega(v),$$

where $d\omega(v)$ is normalised and

$$C(n, p) = \frac{\sqrt{\pi}\,\Gamma\left(\frac{n+p}{2}\right)}{\Gamma\left(\frac{p+1}{2}\right)\Gamma\left(\frac{n}{2}\right)}.$$

The first eigenvalue $\lambda_{p,\Omega}$ of the p-Laplacian satisfies

$$\lambda_{p,\Omega} \ge \left(\frac{p-1}{p}\right)^p \inf\left\{\int_\Omega \frac{|u(x)|^p}{M_p(x)^p}dx : u \in \overset{0}{H}{}^1_p(\Omega), \int_\Omega |u(x)|^p dx = 1\right\}$$

$$\ge \left(\frac{p-1}{p}\right)^p \inf\{M_p(x)^{-p} : x \in \Omega\}.$$

The analogue of Theorem 5.13 is that for any $\rho > 0$,

$$\lambda_{p,\Omega} \geq \left(\frac{p-1}{p}\right)^p \rho^{-p} \left(1 - \sup_{x \in \Omega} \frac{|\Omega \cap B_\rho(x)|}{|B_\rho(x)|}\right), \qquad (5.4.14)$$

which implies, for all $\theta \in (0, 1)$,

$$\lambda_{p,\Omega} \geq \left(\frac{p-1}{p}\right)^p \frac{1-\theta}{\rho_\theta^p}, \text{ where } \rho_\theta := \inf\left\{\rho > 0: \sup_{x \in \Omega} \frac{|\Omega \cap B_\rho(x)|}{|B_\rho(x)|} \leq \theta\right\}.$$
$$(5.4.15)$$

5.5 Extensions of Hardy's Inequality

The refinement

$$\int_0^\pi |u'(x)|^2 dx \geq \frac{1}{4} \int_0^\pi \frac{|u(x)|^2}{\sin^2 x} + \frac{1}{4} \int_0^\pi |u(x)|^2 dx, \quad u \in \overset{0}{H}{}^1(\Omega) \qquad (5.5.1)$$

of the Hardy inequality

$$\int_0^\pi |u'(x)|^2 dx \geq \frac{1}{4} \int_0^\pi \frac{|u(x)|^2}{x^2} dx, \quad u \in \overset{0}{H}{}^1(\Omega)$$

is derived in [87] through knowing about the exact solvability of the differential equation

$$\tau_s y := -\frac{d^2 y}{dx^2} + \frac{s^2 - (1/4)}{\sin^2 x} y = zy, \ z \in \mathbb{C}.$$

Roughly speaking, the non-negativity of the Friedrichs extension associated with the differential expression $\tau_s - (1/4)$, implying the non-negativity of the underlying quadratic form defined on $\overset{0}{H}{}^1(0, \pi)$, yields the refinement. Both constants $1/4$ on the right-hand side of (5.5.1) are shown to be optimal and the inequality is strict in the sense that equality holds if and only if $u = 0$.

In [34] Brezis and Marcus investigated the quantity

$$\lambda^*(\Omega) := \inf_{u \in H_0^1(\Omega)} \frac{\int_\Omega |\nabla u|^2 dx - \frac{1}{4} \int_\Omega |u/\delta|^2 dx}{\int_\Omega |u|^2 dx}$$

for smooth bounded domains Ω. It was shown that the infimum is not achieved, that there are domains for which $\lambda^*(\Omega) < 0$, but for convex domains with C^2 boundary,

$$\lambda^*(\Omega) \geq \frac{1}{4 \operatorname{diam}^2(\Omega)}.$$

Thus with $D(\Omega) := \operatorname{diam}(\Omega)$,

$$\int_\Omega |\nabla u|^2 dx - \frac{1}{4} \int_\Omega |u/\delta|^2 dx \geq \frac{1}{4D(\Omega)^2} \int_\Omega |u|^2 dx, \quad u \in \overset{0}{H}{}^1(\Omega). \qquad (5.5.2)$$

This result generated a great deal of research into obtaining estimates for $\lambda^*(\Omega)$. The problem posed in [34] of whether the diameter $D(\Omega)$ could be replaced in (5.5.2) by a constant multiple of the volume $|\Omega|$ of Ω, i.e.,

$$\lambda^*(\Omega) \geq \alpha \, |\Omega|^{-2/n} \,,$$

was settled in the affirmative in [98] by a method which made use of the mean distance function M_2 and is valid for any domain with non-empty boundary. The approach in [97] was followed in [74] to give the following theorem. Theorem 2.1 in [167] is an L_p analogue for other values of $p > 1$.

Theorem 5.14 *For any $u \in C_0^1(\Omega)$,*

$$\int_\Omega |\nabla u(x)|^2 dx \geq \frac{1}{4} \int_\Omega \frac{|u(x)|^2}{M_2(x)^2} + \frac{3}{2} K(n) \int_\Omega \frac{|u(x)|^2}{|\Omega_x|^{2/n}} \, dx, \qquad (5.5.3)$$

where M_2 is the mean distance function defined in (5.4.1) for $p = 2$, $K(n) := n\left[n^{-1}\omega_n\right]^{2/n}$ and

$$\Omega_x := \{y \in \Omega : x + t(y - x) \in \Omega, \ \forall t \in [0, 1]\},$$

i.e., Ω_x is the set of all $y \in \Omega$ which can be 'seen' from $x \in \Omega$.
If Ω is convex, $\Omega_x = \Omega$ and for any $u \in C_0^1(\Omega)$,

$$\int_\Omega |\nabla u(x)|^2 dx \geq \frac{1}{4} \int_\Omega \frac{|u(x)|^2}{\delta(x)^2} + \frac{3K(n)}{2|\Omega|^{2/n}} \int_\Omega |u(x)|^2 dx. \qquad (5.5.4)$$

Also for Ω convex, Filippas et. al. [77] obtained an estimate

$$\lambda^*(\Omega) \geq \frac{3}{D_{\text{int}}(\Omega)}$$

in terms of the *interior diameter* $D_{\text{int}}(\Omega) := 2 \sup_{x \in \Omega} \delta(x)$. Clearly $D_{\text{int}}(\Omega) \leq D(\Omega)$ and a significant fact is that Ω need not be assumed to be bounded or have finite volume. Following Theorem 3.1 in [77], we now give another form of extension, of the type discussed in [14]. It is assumed that δ is superharmonic in the distributional sense, which we recall from Section 5.3 is satisfied if Ω is convex; in fact the assumption is weaker for $n \geq 3$ but equivalent if $n = 2$. It was also noted in Section 5.2 that δ is superharmonic for a domain Ω with a C^2 boundary 1.1 if and only if Ω is weakly mean convex.

Theorem 5.15 *Let $\Omega \subset \mathbb{R}^n$ be such that $-\Delta\delta \geq 0$ in the distributional sense. Then for any $\alpha > -2$ and all $u \in \overset{0}{H_2^1}(\Omega)$,*

$$\int_\Omega |\nabla\delta(x) \cdot \nabla u(x)|^2 \, dx - \frac{1}{4} \int_\Omega \frac{|u(x)|^2}{\delta(x)^2} \, dx \geq \frac{C_\alpha}{D_{\text{int}}(\Omega)^{\alpha+2}} \int_\Omega \delta(x)^\alpha |u(x)|^2 \, dx$$

$$(5.5.5)$$

with

$$D_\alpha = \begin{cases} 2^\alpha (\alpha + 2)^2, & \alpha \in (-2, -1), \\ 2^\alpha (2\alpha + 3), & \alpha \in [-1, \infty). \end{cases}$$

Under the same conditions, a Hardy–Sobolev–Maz'ya extension of the Hardy inequality was given in [77]:

$$\int_\Omega |\nabla u(x)|^2 \, dx - \frac{1}{4} \int_\Omega \frac{|u(x)|^2}{\delta(x)^2} \, dx \geq C_\Omega \left(\int_\Omega |u(x)|^{2n/(n-2)} dx \right)^{\frac{n-2}{n}}$$

for $n \geq 3$ and all $u \in C_0^\infty(\Omega)$. The problem was posed: can the constant C_Ω be chosen to be independent of Ω? This was settled in [81], where it was also proved that if Ω is a convex domain in $\mathbb{R}^n (n \geq 3)$ and $p \in [2, n)$, there exists a constant $C_{n,p}$, depending only upon n and p, such that

$$\int_\Omega |\nabla u(x)|^p \, dx - \left(\frac{p-1}{p} \right)^p \int_\Omega \frac{|u(x)|^p}{\delta(x)^p} \, dx \geq C_{n,p} \left(\int_\Omega |u(x)|^{pn/(n-p)} dx \right)^{\frac{n-p}{n}} \tag{5.5.6}$$

for all $u \in C_0^\infty(\Omega)$. For Ω the half-space $\mathbb{R}_+^n := \{(x', x_n): x' \in \mathbb{R}^{n-1}, x_n > 0\}$ (and so $\delta(x) = |x_n|$), the case $p = 2$ is proved in [135] and for $2 < p < n$ in [17]; also the sharp value of $C_{3,2}$ is given in [20].

Avkhadiev and Wirths also considered domains Ω which are convex and have finite inradius, and obtained in [6] the following generalisation of the Hardy inequality with weights and sharp constants. An L_p analogue for $p > 2$ is given in [140].

Theorem 5.16 *Let $\Omega \subset \mathbb{R}^n$ be convex with finite inradius $D_{int}(\Omega)$. Then for all $f \in C_0^1(\Omega)$,*

$$\int_\Omega \frac{\nabla f(x)|^2}{\delta(x)^{s-1}} \, dx \geq A \int_\Omega \frac{|f(x)|^2}{\delta(x)^{s+1}} \, dx + \frac{\lambda^2}{D_{int}(\Omega)^q} \int_\Omega \frac{|f(x)|^2}{\delta(x)^{s-q+1}} \, dx, \tag{5.5.7}$$

where A and λ are sharp constants given by

$$A = \frac{s^2 - v^2 q^2}{4} \geq 0, \lambda = \frac{q}{2} \lambda_v (2s/q) > 0,$$

and s, q are positive numbers, $v \in [0, s/q]$ and $z = \lambda_v(s)$ is the Lamb constant defined as the positive root of the Bessel equation $sJ_v(z) + 2J_v'(z) = 0$. In particular, with $s = 1$ and $v = 0$,

$$\int_\Omega |\nabla f(x)|^2 dx \geq \frac{1}{4} \int_\Omega \frac{|f(x)|^2}{\delta(x)^2} \, dx + \frac{\lambda_0^2}{D_{int}(\Omega)^2} \int_\Omega |f(x)|^2 \, dx, \tag{5.5.8}$$

and $\lambda_0 = 0.940...$ is the first zero of $J_0(t) - 2J_1(t)$. The inequality is sharp for $n \geq 1$.

The proof of (5.5.7) is based on the one-dimensional inequality

$$\int_0^1 f'(x)^2 dx > \frac{1}{4} \int_0^1 \frac{f^2(x)}{x^2} dx + \lambda_0^2 \int_0^1 f(x)^2 dx \qquad (5.5.9)$$

for real functions f which are absolutely continuous on $[0, 1]$ and such that $f(0) = 0, f' \in L_2(0, 1)$. The constant λ_0 is shown to be sharp by exhibiting, for each $\varepsilon > 0$, a real function $f \in C_0^1(0, 2)$ which is such that $f'(1) = 0$ and

$$\int_0^1 f'(x)^2 dx < \frac{1}{4} \int_0^1 \frac{f^2(x)}{x^2} dx + \left(\lambda_0^2 + \varepsilon\right) \int_0^1 f(x)^2 dx.$$

Theorem 5.16 is established in [6] by means of inequalities derived from (5.5.9) and the application of an approximation technique of Hadwiger [92] for domains; the same applies to the L_p analogue in [140]. This technique implies, in particular, that for a convex domain $\Omega \subset \mathbb{R}^n$ and any compact set $K \subset \Omega$, there exists a convex n-dimensional polytope Q such that $K \subset \text{int } Q \subset \Omega$. Thus for any $f \in C_0^\infty(\Omega)$, there is a convex n-dimensional polytope Q such that $\text{supp} f \subset \text{int } Q \subset \Omega$. This provides an effective method for establishing many-dimensional inequalities from ones of one dimension, as demonstrated by the many significant contributions made by Avkhadiev and his collaborators.

Brezis–Marcus type inequalities are considered in Section 3.7 of [15] and [124], where assumptions are made on the curvature of the boundary of Ω and δ, rather than the convexity of Ω. Applications to some non-convex domains such as the torus and 1-sheeted hyperboloid follow. For instance, for a ring torus $\Omega \subset \mathbb{R}^3$ with minor ring r and major ring $R \geq 2r$, the ridge $\mathcal{R}(\Omega)$ is closed and of measure zero, $-\Delta\delta > 0$ in $G(\Omega) = \Omega \setminus \mathcal{R}(\Omega)$ and

$$\int_\Omega |\nabla\delta(x) \cdot \nabla f(x)|^p \geq \left(\frac{p-1}{p}\right)^p \int_\Omega \frac{|f(x)|^p}{\delta(x)^p} dx$$
$$+ \left(\frac{p-1}{p}\right)^{p-1} \frac{R-2r}{r(R-r)} \int_\Omega \frac{|f(x)|^p}{\delta(x)^{p-1}} dx$$

$$(5.5.10)$$

for all $f \in C_0^\infty(\Omega)$.

5.6 Discrete Laptev–Weidl Type Inequalities

We present two discrete versions of the Laptev–Weidl inequality (5.1.9); the first is that in [75] on a discretised cylinder $\mathbb{R} \times \mathbb{S}^1$ obtained from the punctured plane $\mathbb{R}^2 \setminus \{0\}$; the second version is the one derived in [91] on the standard lattice $\mathbb{Z}^2$. The following two proofs of (5.1.9) will provide background for the discrete versions considered.

In the Laptev–Weidl inequality

$$\int_{\mathbb{R}^2} |(\nabla + i\mathbf{A})f(x)|^2 \, dx > C \int_{\mathbb{R}^2} \frac{|f(x)|^2}{|x|^2} \, dx, \ C = \min_{k \in \mathbb{Z}} |k - \Psi|^2,$$

given in (5.1.9), the left-hand side is conveniently written in polar co-ordinates as

$$h_A := \int_0^\infty \int_0^\infty \left(\left| \frac{\partial f}{\partial r} \right|^2 + r^{-2} |K_\theta f|^2 \right) r \, dr \, d\theta$$

where $K_\theta := i \frac{\partial}{\partial \theta} + \Psi$ is a self-adjoint operator with domain $H^1(\mathbb{S}^1)$ in $L_1(\mathbb{S}^1)$; it has eigenvalues $\lambda_k = k + \Psi, k \in \mathbb{Z}$ and corresponding eigenvectors

$$\phi_k(\theta) = \frac{1}{\sqrt{2\pi}} \exp(-ik\theta).$$

The sequence $\{\phi_k\}$ is an orthonormal basis of $L_2(\mathbb{S}^1)$ and hence any $f \in L_2(\mathbb{S}^1)$ has the representation

$$f(r, \theta) = \sum_{k \in \mathbb{Z}} f_k(r)\phi_k(\theta),$$

where

$$f_k(r) = \int_0^{2\pi} f(r, \theta)\overline{\phi_k(\theta)} \, d\theta.$$

For any $f \in H^1(\mathbb{S}^1)$,

$$h_A[f] = \sum_{k \in \mathbb{Z}} \int_0^\infty \left(|f_k'(r)|^2 + \frac{\lambda_r^2}{r^2} |f_k(r)|^2 \right) r \, dr$$

and so

$$\int_{\mathbb{R}^2} \frac{|f(x)|^2}{|x|^2} \, dx = \sum_{k \in \mathbb{Z}} \int_0^\infty \frac{|f_k(r)|^2}{r^2} r \, dr$$

$$\leq \sum_{k \in \mathbb{Z}} \frac{1}{\min_{m \in \mathbb{Z}} \lambda_m^2} \int_0^\infty \lambda_k^2 \frac{|f_k(r)|^2}{r^2} r \, dr$$

$$\leq \frac{1}{(\min_{k \in \mathbb{Z}} |k + \Psi|)^2} h_A[f],$$

which proves (5.1.9). We refer to [116], Theorem 3, for a proof that the constant in (5.1.9) is sharp; see also [15], Chapter 5, which also contains background information.

The inequality (5.1.9) can also be proved simply by observing that the punctured plane $\mathbb{R}^2 \setminus \{0\}$ is topologically equivalent to the cylinder $\mathbb{R} \times \mathbb{S}^1$; see [75]. On setting

$$g(u, v) = e^{i \int_0^v \Psi(\theta) d\theta} f(e^u \cos v, e^u \sin v),$$

(5.1.9) with $f \in C_0^\infty (\mathbb{R} \setminus \{0\})$ becomes

$$\int_{-\infty}^{\infty} \int_0^{2\pi} |g(u, v)|^2 \, dv \, du \leq C^{-1} \int_{-\infty}^{\infty} \int_0^{2\pi} \left\{ \left| \frac{\partial g}{\partial u} \right|^2 + \left| \frac{\partial g}{\partial v} \right|^2 \right\} dv \, du, \quad (5.6.1)$$

where $g \in C_0^\infty (\mathbb{R} \times [0, 2\pi])$ and

$$g(u, 2\pi) = e^{2\pi \Psi} g(u, 0), \quad \frac{\partial g}{\partial v}(u, 2\pi) = e^{2\pi \Psi} \frac{\partial g}{\partial v}(u, 0). \quad (5.6.2)$$

On integration by parts, the integral on the right-hand side of (5.6.1) is then seen to equal

$$\int_{-\infty}^{\infty} \int_0^{2\pi} \overline{g} \left(-\frac{\partial^2}{\partial u^2} - \frac{\partial^2}{\partial v^2} \right) g \, dv \, du.$$

The symmetric operator defined by

$$-\frac{\partial^2}{\partial u^2} - \frac{\partial^2}{\partial v^2}$$

for functions in $C_0^\infty (\mathbb{R} \times [0, 2\pi])$ satisfying (5.6.2), has a self-adjoint extension (its Friedrichs extension) with spectrum $[C, \infty)$; this follows by separation of variables, for it decouples into the non-negative self-adjoint operator $-\frac{d^2}{du^2}$ in $L_2(\mathbb{R})$ with spectrum $[0, \infty)$ and the self-adjoint operator $\frac{d^2}{dv^2}$ in $L_2([0, 2\pi])$ with boundary conditions (5.6.2) which has spectrum $\{\lambda \geq 0 : \sqrt{\lambda} = \pm \Psi$ mod $\mathbb{Z}\}$. Hence $C = \min_{k \in \mathbb{Z}} |k - \Psi|^2$ and the inequality (5.1.9) follows.

The form of the Laptev–Weidl inequality in (5.6.1) has a discrete variant which was studied in [75]. This was based on discretising the cylinder $\mathbb{R} \times [0, 2\pi]$ to $\{j/m : j \in \mathbb{Z}\} \times \{2\pi j/n : j \in \{1, 2, ..., n\}\}$; the corresponding point set in the physical plane accumulates at the origin and is exponentially sparse at infinity. On replacing $\partial g/\partial u$ and $\partial g/\partial v$ by $m\{g(j/k) - g(j-1, k)\}$ and $(n/2k)\{g(j, k) - g(j, k-1)\}$ respectively, the discrete analogue of (6.1.6) is

$$\frac{2\pi}{mn} \sum_{j \in \mathbb{Z}} \sum_{k=1}^{n} |g(j, k)|^2$$

$$\leq C^{-1} \frac{2\pi}{mn} \sum_{j \in \mathbb{Z}} \sum_{k=1}^{n} \left\{ m^2 |g(j, k) - g(j-1, k)|^2 + \left(\frac{n}{2\pi} \right)^2 |g(j, k) - g(j, k-1)|^2 \right\}.$$

$$(5.6.3)$$

Denoting by Δ the left-difference operator $\Delta u(l) = u(l) - u(l-1)$ and $\Delta^2 u(l) = \Delta^* \Delta u(l) = u(l+1) + u(l-1) - 2u(l)$, and using the notation

$$\Delta_1 g(j, k) := g(j, k) - g(j-1, k), \quad \Delta_2 g(j, k) := g(j, k) - g(j, k-1),$$

the double sum on the right-hand side of (5.6.3) becomes

$$- \sum_{j \in \mathbb{Z}} \sum_{k \in \Lambda_n} \left\{ m^2 \overline{g(j,k)} \Delta_1^2 g(j,k) + \left(\frac{n}{2\pi} \right)^2 \overline{g(j,k)} \Delta_2^2 g(j,k) \right\}, \qquad (5.6.4)$$

where $\Lambda_n = \{1, 2, ..., n\}$. The following discrete analogue of the Laptev–Weidl inequality is established in [75], Theorem 1.4:

Theorem 5.17 *For all $g \in l_2(\mathbb{Z} \times \Lambda_n)$ and $\Psi \in (0, 1/2]$,*

$$\left\{ \frac{n}{\pi} \sin \left(\frac{\pi}{n} \Psi \right) \right\}^2 \sum_{j \in \mathbb{Z}} \sum_{k \in \Lambda_n} |g(j,k)|^2 \leq D_{mn}[g]$$

$$\leq \left(4m^2 \left\{ \frac{n}{\pi} \sin \left[\frac{\pi}{n} \left(\Psi + \left[\frac{n}{2} \right] \right) \right] \right\}^2 \right) \sum_{j \in \mathbb{Z}} \sum_{k \in \Lambda_n} |g(j,k)|^2 \qquad (5.6.5)$$

where

$$D_{mn}[g] := \sum_{j \in \mathbb{Z}} \sum_{k \in \Lambda_n} \left\{ m^2 |\Delta_1 g(j,k)|^2 + \left(\frac{n}{2\pi} \right)^2 |\Delta_2 g(j,k)|^2 \right\}.$$

Another discrete analogue of a Sobolev-type inequality for Schrödinger operators with Aharonov–Bohm magnetic potential is also derived in [75], Theorem 1.5:

Theorem 5.18 *For all $g \in l_2(\mathbb{Z} \times \Lambda_n)$,*

$$\sup_{j \in \mathbb{Z}} \sum_{k \in \Lambda_n} |g(j,k)|^2 \leq \frac{\pi}{2mn \sin(\Psi \pi / n)} D_{mn}[g].$$

The first step in the proof of the result in [91] is the definition of the discrete Aharonov–Bohm potential. For $k = (k_1, k_2) \in \mathbb{Z}^2$, let $|k|_\infty$ denote the norm $\max\{|k_1|, |k_2|\}$. The circles centred at the origin in this norm are the squares $\mathbb{S}(n) = \{k \in \mathbb{Z}^2 : |k|_\infty = n\}$, with n serving as the radial variable. Denote the phases $\phi_1(k)$ and $\phi_2(k)$ by

$$\phi_1(k) = -\frac{k_2}{8|k|_\infty^2} = +(8n)^{-1}, \quad k_2 = n, \quad -n \leq k_1 \leq n,$$

$$\phi_2(k) = -\frac{k_1}{8|k|_\infty^2} = +(8n)^{-1}, \quad k_1 = -n, \quad -n < k_2 \leq n,$$

$$\phi_1(k) = \frac{k_2}{8|k|_\infty^2} = -(8n)^{-1}, \quad k_2 = -n, \quad -n \leq k_1 \leq n,$$

$$\phi_2(k) = \frac{k_1}{8|k|_\infty^2} = -(8n)^{-1}, \quad k_1 = n, \quad -n \leq k_2 < n.$$

The discrete version of the Aharonov–Bohm potential corresponding to the magnetic flux Ψ is defined as

$$\mathcal{A}_\Psi(k) = -i\,(A_1(k), A_2(k)) = -i\left(1 - e^{2\pi i\Psi\phi_1(k)},\, 1 - e^{2\pi i\Psi\phi_2(k)}\right).$$

Let $e_1 = (1, 0)$ and $e_2 = (0, 1)$. The main result is Theorem 1.1 of [91]:

Theorem 5.19 *For all functions $u\colon \mathbb{Z}^2 \to \mathbb{C}$ decaying sufficiently fast,*

$$\sum_{k\in\mathbb{Z}^2}\sum_{j=1,2} \left|u(k + e_j) - u(k) + iA_j(k)u(k)\right|^2$$

$$\geq 4\sin^2\left(\pi\frac{\mathrm{dist}(\Psi, \mathbb{Z})}{8}\right)\sum_{k\in\mathbb{Z}^2\setminus\{0\}}\frac{|u(k)|^2}{|k|_\infty^2}. \tag{5.6.6}$$

Since $\mathrm{dist}(\Psi, \mathbb{Z}) \leq 1/2$, we have

$$4\sin^2\left(\pi\frac{\mathrm{dist}(\Psi, \mathbb{Z})}{8}\right) \geq 4\left[\pi\frac{\mathrm{dist}(\Psi, \mathbb{Z})}{8}\right]^2\frac{\sin^2(\frac{\pi}{16})}{\left(\frac{\pi}{16}\right)^2}$$

$$= 16\sin^2\left(\frac{\pi}{16}\right)\min_{l\in\mathbb{Z}}|l - \Psi|^2,$$

and (5.6.6) implies

Corollary 5.20 *For all functions $u\colon \mathbb{Z}^2 \to \mathbb{C}$ decaying sufficiently fast,*

$$\sum_{k\in\mathbb{Z}^2}\sum_{j=1,2} \left|u(k + e_j) - u(k) + iA_j(k)u(k)\right|^2$$

$$\geq 16\sin^2\left(\frac{\pi}{16}\right)\min_{k\in\mathbb{Z}^2}|l - \Psi|^2\sum_{k\in\mathbb{Z}^2\setminus\{0\}}\frac{|u(k)|^2}{|k|_\infty^2}. \tag{5.6.7}$$

Note that $16\sin^2\left(\frac{\pi}{16}\right) = 4\left(2 - \sqrt{2 + \sqrt{2}}\right) \sim 0.50896....$

6

Fractional Analogues

6.1 Special Cases and Consequences

6.1.1 Fractional Hardy Inequalities on $\mathbb{R}^n$ and $\mathbb{R}^n_+$

The fractional Hardy inequality on a domain $\Omega \subset \mathbb{R}^n$ with non-empty boundary $\partial\Omega$ has the form

$$\int_{\Omega \times \Omega} \frac{|u(x) - u(y)|^p}{|x - y|^{n+sp}}\, dx\, dx \geq C(s, p, \Omega) \int_{\Omega} \frac{|u(x)|^p}{\delta(x)^s}\, dx, \ u \in C_0^\infty(\Omega), \quad (6.1.1)$$

where $1 < p < \infty$, $0 < s < 1$, $\delta(x) := \inf\{|x - y| : y \in \mathbb{R}^n \setminus \Omega\}$ and $C(s, p, \Omega)$ is a positive constant which is independent of u. The expression on the left-hand side of (6.1.1) is $[u]_{s,p,\Omega}^p$, where $[u]_{s,p,\Omega}$ is the Gagliardo seminorm of u defined in Section 3.1.

We begin our investigation of these inequalities with important special cases on the half-space $\mathbb{R}^n_+ = \{x : x = (x_1, x_2, ..., x_n) \in \mathbb{R}^n, \ x_n > 0\}$ and $\mathbb{R}^n$, and examine significant implications in the latter case for results on the limiting behaviour of fractional inequalities from [23] and [134] discussed in Section 3.2. The first theorem was proved by Bogdan and Dyda in [22] in the case $p = 2$ and extended to all other values of p in [83]. We denote by $\overset{0}{W}{}_p^s(\mathbb{R}^n_+)$ the completion of $C_0^\infty(\mathbb{R}^n_+)$ with respect to the $W_p^s(\mathbb{R}^n_+)$ norm; for $ps < 1$ this coincides with the completion of $C_0^\infty(\overline{\mathbb{R}^n_+})$.

Theorem 6.1 *Let $n \geq 1$, $1 \leq p < \infty$ and $0 < s < 1$ with $ps \neq 1$. Then, for all $u \in \overset{0}{W}{}_p^s(\mathbb{R}^n_+)$,*

$$\int_{\mathbb{R}^n_+ \times \mathbb{R}^n_+} \frac{|u(x) - u(y)|^p}{|x - y|^{n+sp}}\, dx\, dx \geq D_{n,s,p} \int_{\mathbb{R}^n_+} \frac{|u(x)|^p}{|x|^{ps}}\, dx, \quad (6.1.2)$$

with sharp constant

$$D_{n,p,s} := 2\pi^{(n-1)/2} \frac{\Gamma\left((1+ps)/2\right)}{\Gamma\left((n+ps)/2\right)} \int_0^1 |1 - r^{(ps-1)/2}|^p \frac{dr}{(1-r)^{1+ps}}. \quad (6.1.3)$$

If $p = 1$ and $n = 1$, equality holds if and only if u is proportional to a non-increasing function. If $p > 1$ or if $p = 1$ and $n \geq 2$, the inequality is strict for any non-trivial function $u \in \overset{0}{W}{}^s_p(\mathbb{R}^n_+)$.

The theorem follows from the special case $\Omega = \mathbb{R}^n_+$ of an abstract Hardy inequality in [83], Proposition 2.2,

$$E[u] := \int_\Omega \int_\Omega |u(x) - u(y)|^p k(x, y) \, dx \, dy \geq \int_\Omega V(x)|u(x)|^p \, dx, \quad (6.1.4)$$

on compactly supported functions u on $\Omega \subset \mathbb{R}^n$, under the following assumptions. There exists a family of measurable functions k_ε ($\varepsilon > 0$) on $\Omega \times \Omega$ satisfying $k_\varepsilon(x, y) = k_\varepsilon(y, x)$, $0 \leq k_\varepsilon(x, y) \leq k(x, y)$ and $\lim_{\varepsilon \to 0} k_\varepsilon(x, y) = k(x, y)$ for a.e. $x, y \in \Omega$. Moreover, with w a positive, measurable function on Ω, the integrals

$$V_\varepsilon(x) := 2w(x)^{-p+1} \int_\Omega (w(x) - w(y)) \, |w(x) - w(y)|^{p-2} k_\varepsilon(x, y) \, dy$$

are absolutely convergent for a.e. x, belong to $L_{1,loc}(\Omega)$, and $\int V_\varepsilon \phi \, dx \to \int V \phi \, dx$ for any bounded ϕ with compact support in Ω. For the proof of Theorem 6.1, $\Omega = \mathbb{R}^n_+$ and setting $\alpha := (1 - ps)/p$, the following choices are made:

$$w(x) = x_n^{-\alpha}, \quad k(x, y) = |x - y|^{-n-ps}, \quad k_\varepsilon(x, y) = |x - y|^{-n-ps} \chi_{|x_n - y_n|}.$$

Then [82], Lemma 3.1 gives that $V(x) = D_{n,p,s} x_n^{-ps}$ and hence

$$2 \lim_{\varepsilon \to 0} \int_{||x| - |y|| > \varepsilon} (w(x) - w(y)) \, |w(x) - w(y)|^{p-2} k(x, y) \, dy = \frac{D_{n,p,s}}{|x|^{ps}} w(x)^{p-1}. \quad (6.1.5)$$

Therefore (6.1.2) is established. We refer to the proof of Theorem 1.1 in [83] for showing that the constant $D_{n,p,s}$ in (6.1.3) is optimal and also for details on the remainder of Theorem 6.1.

The approach sketched above for establishing Theorem 6.1 in [83], based on (6.1.4) with $\Omega = \mathbb{R}^n_+$, is used for $\Omega = \mathbb{R}^n$ in [82], and, in fact, will be used for a general domain Ω in Section 6.2. The choices

$$w(x) = |x|^{-\alpha}, \quad k(x, y) = |x - y|^{-n-ps}, \quad V(x) = C(n, s, p)|x|^{-ps}$$

yield the following modification in [82]:

Theorem 6.2   *Let $n \geq 1$ and $0 < s < 1$. Then for all $u \in \overset{0}{W}_p^s(\mathbb{R}^n) = W_p^s(\mathbb{R}^n)$*

(see (3.2.2)) if $1 \leq p < n/s$, and for all $u \in \overset{0}{W}_p^s(\mathbb{R}^n \setminus \{0\})$ if $p > n/s$,

$$\int_{\mathbb{R}^n} \int_{\mathbb{R}^n} \frac{|u(x) - u(y)|^p}{|x-y|^{n+sp}} \, dx \, dx \geq C(n,s,p) \int_{\mathbb{R}^n} \frac{|u(x)|^p}{|x|^{ps}} \, dx, \qquad (6.1.6)$$

where

$$C_{n,s,p} := 2 \int_0^1 r^{ps-1} |1 - r^{(n-ps)/p}|^p \Phi_{n,s,p}(r) \, dr, \qquad (6.1.7)$$

and

$$\Phi_{n,s,p}(r) := \omega_{n-2} \int_{-1}^1 \frac{(1-t^2)^{(n-3)/2}}{(1-2rt+r^2)^{(n+ps)/2}} \, dt, \ n \geq 2,$$

$$\Phi_{1,s,p}(r) := \left(\frac{1}{(1-r)^{1+ps}} + \frac{1}{(1+r)^{1+ps}} \right), \ n-1.$$

The constant $C_{n,s,p}$ is optimal. If $p = 1$, equality holds if and only if u is proportional to a symmetric decreasing function. If $p > 1$, the inequality is strict for any non-trivial function $u \in W_p^s(\mathbb{R}^n)$ or $\overset{0}{W}_p^s(\mathbb{R}^n \setminus \{0\})$, respectively.

6.1.2 The Limiting Cases of $s \to 0+$ and $s \to 1-$

It is proved in [134], Theorem 2, that for $n \geq 1$, $0 < s < 1$, $1 \leq p < n/s$ and $u \in W_p^s(\mathbb{R}^n)$,

$$\int_{\mathbb{R}^n} \int_{\mathbb{R}^n} \frac{|u(x) - u(y)|^p}{|x-y|^{n+sp}} \, dx \, dy \geq c(n,p) \frac{(n-sp)^p}{s(1-s)} \int_{\mathbb{R}^n} \frac{|u(x)|^p}{|x|^{ps}} \, dx \qquad (6.1.8)$$

for some constant $c(n,p)$ which depends only on n and p. Since $C(n,s,p)$ in Theorem 6.2 is optimal, it follows that

$$c(n,p) \frac{(n-sp)^p}{s(1-s)} \leq C(n,s,p). \qquad (6.1.9)$$

There are related consequences of Theorem 3.19, where we saw that for $p \in (1, \infty)$ and $u \in W_p^s(\mathbb{R}^n)$, there exists a positive constant $K(p,n)$ such that

$$\lim_{s \to 1-} (1-s) \int_{\mathbb{R}^n} \int_{\mathbb{R}^n} \frac{|u(x) - u(y)|^p}{|x-y|^{n+sp}} \, dx \, dy = \frac{K(p,n)}{p} \int_{\mathbb{R}^n} |\nabla u(x)|^p \, dx$$

and hence by Hardy's inequality,

$$\lim_{s \to 1-} (1-s) \int_{\mathbb{R}^n} \int_{\mathbb{R}^n} \frac{|u(x) - u(y)|^p}{|x-y|^{n+sp}} \, dx \, dy \geq \frac{K(p,n)}{p} \left(\frac{p-1}{p} \right)^p \int_{\mathbb{R}^n} \frac{|u(x)|^p}{|x|^p} \, dx. \qquad (6.1.10)$$

Also, for $u \in \bigcup_{0<s<1} W_p^s(\mathbb{R}^n)$, there exists a positive constant $C'(n, p) \approx p^{-1}n$ such that

$$\lim_{s\to 0+} s \int_{\mathbb{R}^n} \int_{\mathbb{R}^n} \frac{|u(x) - u(y)|^p}{|x-y|^{n+sp}} \, dx \, dy = C'(n, p) \int_{\mathbb{R}^n} |u(x)|^p dx; \qquad (6.1.11)$$

see Remark 3.20.

It is fitting to recall here that for $ps < 1$, the Sobolev embedding theorem asserts that $W_p^s(\mathbb{R}^n) \hookrightarrow L_{p*}(\mathbb{R}^n)$, where $p^* = np/(n-ps)$, and

$$\left(\int_{\mathbb{R}^n} \int_{\mathbb{R}^n} \frac{|u(x)-u(y)|^p}{|x-y|^{n+sp}} \, dx \, dy \right)^{1/p} \geq S_{n,s,p} \int_{\mathbb{R}^n} |u(x)|^{p*} dx. \qquad (6.1.12)$$

The optimal values of the constants $S_{n,s,p}$ are not known. Estimates are given in [23] which reflect the correct behaviour as s tends to 1; in [134], Theorem 1, the sharp constant is shown to satisfy

$$S_{n,s,p} \geq c(n,p) \frac{(n-ps)^{p-1}}{s(1-s)} \qquad (6.1.13)$$

for some positive constant $c(n, p)$, and for $u \in \bigcup_{0<s<1} W_p^s(\mathbb{R}^n)$, the asymptotic result

$$\lim_{s\to 0+} s \int_{\mathbb{R}^n} \int_{\mathbb{R}^n} \frac{|u(x)-u(y)|^p}{|x-y|^{n+sp}} \, dx \, dy = 2p^{-1}\omega_{n-1} \int_{\mathbb{R}^n} |u(x)|^p \, dx \qquad (6.1.14)$$

is proved in [134] Theorem 3. The following asymptotic result when $s \to 1-$ is established in [150] and [30] for all $u \in C_0^\infty(\Omega)$ and $\Omega \subset \mathbb{R}^n$ convex:

$$\lim_{s\to 1-} (1-s) \int_{\mathbb{R}^n} \int_{\mathbb{R}^n} \frac{|u(x)-u(y)|^p}{|x-y|^{n+sp}} \, dx \, dy = \alpha_{n,p} \int_{\Omega} |\nabla u(x)|^p \, dx, \qquad (6.1.15)$$

where, with $\mathbf{e_1} = (1, 0, 0, ..., 0)$,

$$\alpha_{n,p} = \frac{1}{p} \int_{\mathbb{S}^{n-1}} |\langle \sigma \cdot \mathbf{e_1} \rangle|^p d\sigma = \frac{\Gamma\left(\frac{n}{2}\right)\Gamma\left(\frac{p+1}{2}\right)}{\pi^{1/2}\Gamma\left(\frac{n+p}{2}\right)}.$$

Thus an extension of Corollary 3.20 is achieved which allows for the inequality on a convex subset Ω of $\mathbb{R}^n$.

In [145], Peetre proved that the standard Sobolev embedding $W_p^s(\mathbb{R}^n) \hookrightarrow L_{p*}(\mathbb{R}^n)$ can be refined to $W_p^s(\mathbb{R}^n) \hookrightarrow L_{p*,p}(\mathbb{R}^n)$. The following sharp inequality associated with this embedding is given in [82], Theorem 4.1:

Theorem 6.3 *Let $n \in \mathbb{N}$, $0 < s < 1$, $1 \leq p < n/s$ and $p^* = np/(n-ps)$. Then $W_p^s(\mathbb{R}^n) \hookrightarrow L_{p*,p}(\mathbb{R}^n)$ and*

$$\|u\|_{p*,p} \leq \left(\frac{n}{\omega_{n-1}} \right)^{s/n} C_{n,p,s}^{-1/p} \left(\int_{\mathbb{R}^n} \int_{\mathbb{R}^n} \frac{|u(x)-u(y)|^p}{|x-y|^{n+sp}} \, dx \, dy \right)^{1/p} \qquad (6.1.16)$$

for any $u \in W_p^s(\mathbb{R}^n)$ with $C_{n,p,s}$ from (6.1.7). The constant is sharp. For $p = 1$, equality holds if u is proportional to a non-negative function v such that the level sets $\{v > \tau\}$ are balls for a.e. τ. For $p > 1$ the inequality is strict for any non-trivial u.

In [21], Proposition 4.2, it is shown that if $1 \le p < r$ and $0 < q \le r \le \infty$, then

$$\|u\|_{q,r} \le \left(\frac{q}{p}\right)^{\frac{(r-p)}{rp}} \|u\|_{q,p}.$$

It follows that we have

Corollary 6.4 *Let $n \ge 1$, $0 < s < 1$, $1 \le p < n/s$, $p^* = np/(n - ps) \le r \le \infty$ and $p < r$. Then $W_p^s(\mathbb{R}^n) \hookrightarrow L_{p^*,r}(\mathbb{R}^n)$ and*

$$\|u\|_{p^*,r} \le \left(\frac{p*}{p}\right)^{\frac{r-p}{rp}} \left(\frac{n}{\omega_{n-1}}\right)^{s/n} C_{n,p,s}^{-1/p} \left(\int_{\mathbb{R}^n} \int_{\mathbb{R}^n} \frac{|u(x) - u(y)|^p}{|x - y|^{n+sp}} \, dx \, dy\right)^{1/p}.$$
$$(6.1.17)$$

The choice $r = p^*$ gives the Sobolev inequality

$$\int_{\mathbb{R}^n} \int_{\mathbb{R}^n} \frac{|u(x) - u(y)|^p}{|x - y|^{n+sp}} \, dx \, dy \ge S_{n,s,p} \|u\|_{p^*}^p$$

where

$$S_{n,s,p} = \left(\frac{p}{p*}\right)^{ps/n} \left(\frac{\omega_{n-1}}{n}\right)^{ps/n} C_{n,s,p}. \qquad (6.1.18)$$

The asymptotic inequality (6.1.11) is then recovered from (6.1.9).

It is proved in [82], Lemma 4.3, that for $0 < s \le 1$ and $1 \le p < n/s$, any non-negative symmetric decreasing function u on $\mathbb{R}^n$ satisfies

$$\|u\|_{p^*,p} = \left(\frac{n}{\omega_{n-1}}\right)^{s/n} \left(\int_{\mathbb{R}^n} \frac{u(x)^p}{|x|^{ps}} \, dx\right)^{1/p}. \qquad (6.1.19)$$

This establishes the link between Theorem 6.3 and the sharp inequality (6.1.6).

The 'local' analogue of (6.1.16) with $s = 1$ is

$$\|u\|_{p^*,p} \le \left(\frac{n}{\omega_{n-1}}\right)^{1/n} \frac{p}{n-p} \left(\int_{\mathbb{R}^n} |\nabla u(x)|^p dx\right)^{1/p} \qquad (6.1.20)$$

for $n \ge 2$, $1 \le p < n$ and $p^* = np/(n - p)$. This is the inequality (5.1.4) with the optimal constant proved by Alvino in [10].

A fractional Hardy inequality which focuses on the dependence of the constant on s is featured in the following theorem in [25] in which the correct asymptotic behaviour with respect to s as $s \to 0+$ and $s \to 1-$ is exhibited in accordance with (6.1.14) and (6.1.15). Note that the inequality (6.1.21) is not of the type (6.1.1) since the integration on the left-hand side of (6.1.21) is over the whole of $\mathbb{R}^n \times \mathbb{R}^n$ rather than $\Omega \times \Omega$, as in (6.1.1). The result should be compared with Theorem 6.8, which has the same domain of integration on both sides, gives the right asymptotic behaviour as $s \to 1-$ and $s \to 0+$, and has the best constant, but at the expense of requiring the condition $1/p < s < 1$.

Theorem 6.5 *Let* $1 < p < \infty$ *and* $0 < s < 1$, *and let* Ω *be a proper open convex subset of* $\mathbb{R}^n, n \geq 1$. *Then for all* $u \in C_0^\infty(\Omega)$ *and with* $\delta(x) :=$ $\inf\{|x - y| : y \notin \Omega\}$, *there exists a positive constant* $C(n, p)$ *such that*

$$\int_{\mathbb{R}^n} \int_{\mathbb{R}^n} \frac{|u(x) - u(y)|^p}{|x - y|^{n+sp}} \, dx \, dy \geq C(n, p) \frac{1}{s(1-s)} \int_\Omega \frac{|u(x)|^p}{\delta(x)^{ps}} \, dx. \quad (6.1.21)$$

This is a nonlocal analogue of the inequality in Section 5.3 for the local Hardy inequality on convex domains. The proof is based on the property that for $0 < s < 1$ and $1 < p < \infty$, δ^s is (s, p)-*locally weakly superharmonic* in the following weak sense:

$$\int_{\mathbb{R}^n} \int_{\mathbb{R}^n} \frac{|\delta(x)^s - \delta(y)^s|^{p-2}(\delta(x)^s - \delta(y)^s)(\phi(x) - \phi(y))}{|x - y|^{n+sp}} \, dx \, dy \geq 0 \quad (6.1.22)$$

for all non-negative $\phi \in W_s^p(\Omega)$ with compact support in Ω. This property is written as $(-\Delta_p)^s \delta^s \geq 0$, where $(-\Delta_p)^s$ is the fractional p-Laplacian of order s; see (3.4.8) for more details on the significance of this property. The pivotal result in [25], Proposition 3.2, is that if Ω is an open, bounded, convex subset of $\mathbb{R}^n$, then δ^s is locally weakly (s, p)-superharmonic. An interesting preliminary result from [102] (and from [38] when $p = 2$) is that for $\Omega := \mathbb{R}^n_+$, δ^s is locally weakly (s, p)-harmonic, i.e., δ^s and $-\delta^s$ are both locally weakly (s, p)-superharmonic.

The proof of Theorem 6.5 uses a Moser-type argument which is similar to that in Section 5.3, this time choosing

$$\phi = \frac{|u|^p}{(\delta^s + \varepsilon)^{p-1}}$$

where $u \in C_0^\infty(\Omega)$ and $\varepsilon > 0$; in [25], Lemma 2.4, this choice is proved to be admissible, i.e., $\phi \in W_s^p(\Omega)$. Another important role in the proof is played by the following fractional counterpart from [25], Proposition 2.5, of the property that $|\nabla \delta| = 1$ a.e. in Ω.

Proposition 6.6 *Let* $1 < p < \infty$, $0 < s < 1$, *and let* Ω *be an open, bounded, convex subset of* $\mathbb{R}^n$. *Then*

$$\int_{y \in \Omega : \, \delta(y) \leq \delta(x)} \frac{|\delta(x) - \delta(y)|^p}{|x - y|^{n+sp}} \, dy \geq \frac{C_1}{1-s} \delta(x)^{p(1-s)}$$

for a.e. $x \in \Omega$, where

$$C_1 = C_1(n, p) = \frac{1}{p} \sup_{0<\sigma<1} \left[\sigma^p \mathcal{H}^{n-1} \left(\{ \omega \in \mathbb{S}^{n-1} : \langle \omega, e_1 \rangle > \sigma \} \right) \right];$$

$n-1$ *is the Hausdorff dimension of the exhibited set and $\mathcal{H}^{n-1}(\cdot)$ is its $(n-1)$-dimensional Hausdorff measure.*

An analysis of the dependence of (6.1.1) on Ω, and the permissible values of p and s, is given in the following theorem from [55]:

Theorem 6.7 *The inequality (6.1.1) holds in each of the following cases:*

1. *Ω is a bounded Lipschitz domain and $ps > 1$;*
2. *$\Omega = \mathbb{R}^n \setminus K$, where K is a bounded Lipschitz domain, $ps \neq 1$, $ps \neq n$;*
3. *Ω is a domain above the graph of a Lipschitz function $\mathbb{R}^{n-1} \to \mathbb{R}^n$ and $ps \neq 1$;*
4. *Ω is the complement of a point and $ps \neq n$.*

Furthermore, (6.1.1) does not hold if

1. *Ω is a bounded Lipschitz domain and $ps \leq 1$, $s < 1$;*
2. *Ω is the complement of a compact set and $n = ps$, $s < p$.*

A consequence is that in the integral on the left-hand side of (6.1.21), $\mathbb{R}^n \times \mathbb{R}^n$ cannot be replaced by $\Omega \times \Omega$ whenever Ω is bounded if $ps \leq 1$. It is observed in [55] (see also [44]) that if Ω is a bounded Lipschitz domain and $ps \leq 1$,

$$\int_\Omega \frac{|u(x)|^p}{\delta(x)^{ps}} \, dx \leq C \left(\int_\Omega \int_\Omega \frac{|u(x) - u(y)|^p}{|x - y|^{n+ps}} \, dx + \int_\Omega |u(x)|^p \, dx \right) \quad (6.1.23)$$

for all $u \in C_0^\infty(\Omega)$, the inequality being false without the final term even for domains Ω with a C^∞ boundary.

In [23], Theorem 1, it is proved for Ω a cube in $\mathbb{R}^n$ ($n > 1$), $0 < s < 1$, $p > 1$, $ps < n$, $\frac{1}{q} = \frac{1}{p} - \frac{s}{n}$ and $u \in W_s^p(\Omega)$, that

$$\int_\Omega \int_\Omega \frac{|u(x) - u(y)|^p}{|x - y|^{n+ps}} \, dx \geq C(n)(n/q)^p (1 - s)^{-1} \int_\Omega |u(x) - u_\Omega|^q \, dx, \quad (6.1.24)$$

where $u_\Omega := (1/|\Omega|) \int_\Omega u(x) \, dx$ and $C(n)$ depends only on n. Related work on inequalities of fractional Hardy and Poincaré type may be found in [58], [62], [99], [100] and [101].

6.2 General Domains

This section is mainly devoted to the fractional analogue of the local Hardy-type inequality in Theorem 5.10 on a general domain Ω in $\mathbb{R}^n (n \geq 2)$ with non-empty boundary, involving a mean distance function $M_{p,\Omega}$. The appropriate mean distance function now depends on $s \in (0, 1)$ and is defined by

$$\frac{1}{M_{s,p}(x)^{ps}} := \frac{\pi^{1/2}\Gamma\left(\frac{n+ps}{2}\right)}{\Gamma\left(\frac{1+ps}{2}\right)\Gamma\left(\frac{n}{2}\right)} \int_{\mathbb{S}^{n-1}} \frac{1}{\delta_v^{ps}(x)}\, d\omega(v), \qquad (6.2.1)$$

where $1/p < s < 1$,

$$\tau_v(x) = \min\{t > 0 : x + tv \notin \Omega\}, \quad \delta_v(x) = \min\{\tau_v(x), \tau_{-v}(x)\}$$

and the surface measure ω on $\mathbb{S}^{n-1}$ is normalised, i.e., $\int_{\mathbb{S}^{n-1}} d\omega(v) = 1$. If Ω is convex, then, as in Theorem 5.12, $M_{s,p}(x) \leq \delta(x)$.

Loss and Sloane show in [130] that Theorem 6.1 continues to hold with the same sharp constant for any convex domain Ω. Their proof makes use of a mean distance function and it is this which is the basis of this section.

Theorem 6.8 *Let Ω be an open subset of $\mathbb{R}^n$ with non-empty boundary, let $p \in (1, \infty)$ and $s \in (1/p, 1)$. Then for all $f \in C_0^\infty(\Omega)$,*

$$\int_\Omega \int_\Omega \frac{|f(x) - f(y)|^p}{|x - y|^{n+ps}}\, dx\, dy \geq D_{n,p,ps} \int_\Omega \frac{|f(x)|^p}{M_{s,p}(x)^{ps}}\, dx, \qquad (6.2.2)$$

where

$$D_{n,p,ps} := \frac{\pi^{(n-1)/2}\Gamma\left(\frac{1+ps}{2}\right)}{\Gamma\left(\frac{n+ps}{2}\right)} D_{1,p,ps} \qquad (6.2.3)$$

and

$$D_{1,p,ps} := 2\int_0^1 \frac{|1 - r^{(ps-1)/p}|}{(1 - r)^{1+ps}}\, dr. \qquad (6.2.4)$$

For Ω convex,

$$\int_\Omega \int_\Omega \frac{|f(x) - f(y)|^p}{|x - y|^{n+ps}}\, dx\, dy \geq D_{n,p,ps} \int_\Omega \frac{|f(x)|^p}{\delta(x)^{ps}}\, dx. \qquad (6.2.5)$$

The inequality (6.2.5) with some constant $C(n, p)$ continues to hold for $0 < p \leq 1$, but the optimal value of the constant is not known, see [55].

The proof depends on the following one-dimensional inequality.

Lemma 6.9 *Let $1 < p < \infty$, $1 < p < s < 1$ and $f \in C_0^\infty(a, b)$. Then*

$$\int_a^b \int_a^b \frac{|f(x) - f(y)|^p}{|x - y|^{1+ps}}\, dx\, dy \geq D_{1,p,ps} \int_a^b \frac{|f(x)|^p}{\min\{(x - a), (b - x)\}^{ps}}\, dx. \qquad (6.2.6)$$

Moreover, if $J \subset (a, b)$ is an open set and $f \in C_0^\infty(J)$, then for all $f \in C_0^\infty(J)$,

$$\int_J \int_J \frac{|f(x) - f(y)|^p}{|x - y|^{1+ps}} \, dx \, dy \geq D_{1,p,ps} \int_J \frac{|f(x)|^p}{\delta_J(x)^{ps}} \, dx; \qquad (6.2.7)$$

J is a countable union of disjoint intervals I_k, and so for $x \in J$ there is a unique interval I_k containing x and then $\delta_J(x) = \delta_{I_k(x)} := \inf\{|t| : x + t \notin I_k\}$.

Proof From [83], Proposition 2.2 and Lemma 2.4,

$$\int_a^b \int_a^b \frac{|f(x) - f(y)|^p}{|x - y|^{1+\alpha}} \, dx \, dy \geq \int_a^b V(x)|f(x)|^p dx,$$

where, with $w(x) = \delta_{(a,b)}(x)^{-(1-\alpha)/p}$, $\alpha = ps$,

$$V(x) = \frac{2}{w(x)^{p-1}} \int_a^\infty (w(x) - w(y)) |w(x) - w(y)|^{p-2} \frac{dy}{|x - y|^{1+\alpha}}. \qquad (6.2.8)$$

It is necessary to prove that $V(x) \geq \frac{1}{\delta(x)^\alpha} D_{1,p,\alpha}$.

Let

$$\begin{aligned} I(x) :=\ & 2 \int_0^\infty (w(x) - w(y)) |w(x) - w(y)|^{p-2} \frac{dy}{|x-y|^{1+\alpha}} \\ =\ & 2 \left(\int_0^x + \int_x^\infty \right) (w(x) - w(y)) |w(x) - w(y)|^{p-2} \frac{dy}{|x-y|^{1+\alpha}} \qquad (6.2.9) \\ =:\ & 2 \left(I_1(x) + I_2(x) \right), \end{aligned}$$

where

$$w(x) = x^{-(1-\alpha)/p}, \ 1 < p < \infty, \ 1 < \alpha < 2.$$

Note that, to be precise, the integrals in (6.2.8), like the integral in (6.2.7), are principal values, being over $(0, \varepsilon)$ and (ε, ∞) and the limit as $\varepsilon \to 0+$ taken. Then, on putting $y = tx$,

$$\begin{aligned} I_1(x) &= x^{\frac{(\alpha-1)}{p}(p-1)-\alpha} \int_0^1 \left(1 - t^{\frac{\alpha-1}{p}} \right) \left| \left(1 - t^{\frac{\alpha-1}{p}} \right) \right|^{p-2} \frac{dt}{|1-t|^{1+\alpha}} \\ &= x^{\frac{(\alpha-1)}{p}(p-1)-\alpha} \int_0^1 \left| \left(1 - t^{\frac{\alpha-1}{p}} \right) \right|^{p-1} \frac{dt}{|1-t|^{1+\alpha}}. \end{aligned}$$

On putting $y = x/t$,

$$\begin{aligned} I_2(x) &= -x^{\frac{(\alpha-1)}{p}(p-1)-\alpha} \int_1^0 \left(1 - t^{-(\alpha-1)/p} \right) \left| 1 - t^{-(\alpha-1)/p} \right|^{p-2} \frac{dt}{t^2 |1 - 1/t|^{1+\alpha}} \\ &= -x^{\frac{(\alpha-1)}{p}(p-1)-\alpha} \int_1^0 \left(t^{(\alpha-1)/p} - 1 \right) \left| t^{(\alpha-1)/p} - 1 \right|^{p-2} \frac{t^{(\alpha-1)/p} dt}{|1-t|^{1+\alpha}} \\ &= -x^{\frac{(\alpha-1)}{p}(p-1)-\alpha} \int_0^1 \left| 1 - t^{\frac{\alpha-1}{p}} \right|^{p-1} \frac{t^{(\alpha-1)/p} dt}{|1-t|^{1+\alpha}}. \end{aligned}$$

Thus

$$I(x) = 2x^{\frac{(\alpha-1)}{p}(p-1)-\alpha} \int_0^1 \left|1 - t^{(\alpha-1)/p}\right|^{p-1} \left(1 - t^{(\alpha-1)/p}\right) \frac{dt}{|1-t|^{1+\alpha}}$$

$$= 2x^{\frac{(\alpha-1)}{p}(p-1)-\alpha} \int_0^1 \left|1 - t^{(\alpha-1)/p}\right|^p \frac{dt}{|1-t|^{1+\alpha}}$$

$$= x^{\frac{(\alpha-1)}{p}(p-1)-\alpha} \mathcal{D}_{1,p,\alpha}; \tag{6.2.10}$$

also

$$I_1(x) \geq x^{\frac{(\alpha-1)}{p}(p-1)-\alpha} \mathcal{D}_{1,p,\alpha}. \tag{6.2.11}$$

From (6.2.7) with $w(x) = \delta_{(a,b)}(x)^{(\alpha-1)/p}$ and $c = (1/2)(a+b)$,

$$w(x)^{p-1} V(x) = 2 \int_a^c (w(x) - w(y)) |w(x) - w(y)|^{p-2} \frac{dy}{|x-y|^{(1+\alpha)}} \tag{6.2.12}$$

for $a < x < c$ and

$$w(x)^{p-1} V(x) = 2 \int_c^b (w(x) - w(y)) |w(x) - w(y)|^{p-2} \frac{dy}{|x-y|^{(1+\alpha)}} \tag{6.2.13}$$

for $c < x < b$. Similar calculations to those which yield (6.2.9) and (6.2.10) now give the following: for $a < x < c$,

$$\frac{2}{w(x)^{p-1}} \int_a^\infty (w(x) - w(y)) |w(x) - w(y)|^{p-2} \frac{dy}{|x-y|^{(1+\alpha)}}$$

$$= \frac{1}{(x-a)^\alpha} \mathcal{D}_{1,p,\alpha} \tag{6.2.14}$$

and

$$V(x) \geq \frac{1}{(x-a)^\alpha} \mathcal{D}_{1,p,\alpha}, \tag{6.2.15}$$

while for $c < x < b$,

$$\frac{2}{w(x)^{p-1}} \int_{-\infty}^b (w(x) - w(y)) |w(x) - w(y)|^{p-2} \frac{dy}{|x-y|^{(1+\alpha)}}$$

$$= \frac{1}{(b-x)^\alpha} \mathcal{D}_{1,p,\alpha} \tag{6.2.16}$$

and

$$V(x) \geq \frac{1}{(b-x)^\alpha} \mathcal{D}_{1,p,\alpha}. \tag{6.2.17}$$

Therefore, $V(x) \geq \frac{1}{\delta(x)^\alpha} \mathcal{D}_{1,p,\alpha}$ and (6.2.5) is proved.

Since an open subset J is the countable union of disjoint intervals I_k, we have from (6.2.5)

$$
\int_J \int_J \frac{|f(x) - f(y)|^p}{|x-y|^{1+ps}}\, dx\, dy \geq \sum_{k=1}^{\infty} \int_{I_k} \int_{I_k} \frac{|f(x) - f(y)|^p}{|x-y|^{1+ps}}\, dx\, dy
$$

$$
\geq \sum_{k=1}^{\infty} D_{1,p,ps} \int_{I_k} \frac{|f(x)|^p}{\delta_{I_k}(x)^{ps}}\, dx
$$

$$
\geq D_{1,p,ps} \int_J \frac{|f(x)|^p}{\delta_J(x)^{ps}}\, dx.
$$

$\square$

We now quote Lemma 2.4 in [130] which leads to the application of Lemma 6.9 and proof of Theorem 6.8.

Lemma 6.10 *Let Ω be a domain in $\mathbb{R}^n$. Then for all $f \in C_0^{\infty}(\Omega)$,*

$$
\int_\Omega \int_\Omega \frac{|f(x) - f(y)|^p}{|x-y|^{1+ps}}\, dx\, dy
$$
$$
= \frac{\omega_{n-1}}{2} \int_{\mathbb{S}^{n-1}} d\omega \int_{x:\, x\cdot\omega=0} dL_\omega(x) \int_{x+s\omega\in\Omega} ds \int_{x+t\omega\in\Omega} \frac{|f(x+s\omega) - f(x+t\omega)|^p}{|s-t|^{1+ps}}\, dt
$$
$$
(6.2.18)
$$

where $\mathcal{L}_\omega$ denotes the $(n-1)$-dimensional Lebesgue measure on the plane $x \cdot \omega = 0$; recall that the measure $d\omega$ on $\mathbb{S}^{n-1}$ is normalised.

Proof Let

$$
I_\Omega(f) := \int_\Omega \int_\Omega \frac{|f(x) - f(y)|^p}{|x-y|^{1+ps}}\, dx\, dy = \int_\Omega dx \int_{x+z\in\Omega} \frac{|f(x) - f(x+z)|^p}{|z|^{n+ps}}\, dz.
$$

On using polar co-ordinates $z = r\omega$, we obtain

$$
I_\Omega(f) = \omega_{n-1} \int_\Omega dx \int_{\mathbb{S}^{n-1}} d\omega \int_{x+r\omega\in\Omega,\, r>0} \frac{|f(x) - f(x+r\omega)|^p}{r^{1+ps}}\, dr
$$
$$
= \frac{1}{2}\omega_{n-1} \int_{\mathbb{S}^{n-1}} d\omega \int_\Omega dx \int_{x+h\omega\in\Omega} \frac{|f(x) - f(x+h\omega)|^p}{|h|^{1+ps}}\, dh.
$$

The domain of integration $\{x + h\omega \in \Omega\}$ in the innermost integral is the line $x + h\omega$ intersected with Ω. On splitting the variable x into components perpendicular to ω and parallel to ω, i.e., replacing x by $x + l\omega$, where $x \cdot \omega = 0$, we derive

$$
\frac{\omega_{n-1}}{2} \int_{\mathbb{S}^{n-1}} d\omega \int_{x:\, x\cdot\omega=0} dL_\omega(x) \int_{x+l\omega\in\Omega} dl \int_{x+(l+h)\omega\in\Omega} \frac{|f(x+l\omega) - f(x+(l+h)\omega)|^p}{|h|^{1+ps}}\, dh.
$$

The lemma follows by the variable change $t = l + h$.

□

Proof of Theorem 6.8 By Lemma 6.10 and (6.2.6),

$$
\begin{aligned}
&\int_\Omega \int_\Omega \frac{|f(x) - f(y)|^p}{|x - y|^{1+ps}} \, dx \, dy \\
&= \frac{1}{2} \omega_{n-1} \int_{\mathbb{S}^{n-1}} d\omega \int_{x:\, x\cdot\omega=0} dL_\omega(x) \int_{x+l\omega\in\Omega} ds \int_{x+t\omega\in\Omega} \frac{|f(x+s\omega)-f(x+t\omega)|^p}{|s-t|^{1+ps}} \, dt \\
&\geq \frac{1}{2} \omega_{n-1} D_{1,p,ps} \int_{\mathbb{S}^{n-1}} d\omega \int_{x:\, x\cdot\omega=0} d\mathcal{L}_\omega(x) \int_{x+l\omega\in\Omega} \frac{|f(x+l\omega)|^p}{\delta_\omega(x+l\omega)^{ps}} \, dl \\
&= \frac{1}{2} \omega_{n-1} D_{1,p,ps} \int_{\mathbb{S}^{n-1}} d\omega \int_\Omega \frac{|f(x)|^p}{\delta_\omega(x)^{ps}} \, dx \\
&= D_{n,p,ps} \int_\Omega \frac{|f(x)|^p}{M_{s,p}(x)^{ps}} \, dx,
\end{aligned}
$$

since

$$
\frac{\omega_{n-1}}{2} \frac{\Gamma(\frac{n}{2})\Gamma(\frac{1+ps}{2})}{\sqrt{\pi}\,\Gamma(\frac{n+ps}{2})} D_{1,p,ps} = D_{n,p,ps}.
$$

In the case $\Omega = \mathbb{R}^n_+$, the constant $D_{n,p,ps}$ was proved to be best possible in [22] for $p = 2$ and in [83] for the other values of p, by constructing a sequence of trial functions. For a general convex Ω, these trial functions are transplanted to Ω near a tangent hyperplane, following the proof of Theorem 5 in [132].

The theorem is therefore proved. □

Remark 6.11

In the case $p = 2$, it is proved in the appendix of [22] that

$$
D_{1,2,2s} = \frac{1}{s}\left\{ \frac{2^{-2s}}{\sqrt{\pi}} \Gamma\left(\frac{1+2s}{2}\right) \Gamma(1-s) - 1 \right\}. \tag{6.2.19}
$$

Also, for $p = 2$ an improvement of (6.2.2) is established in [130], Theorem 1.1, namely,

$$
\int_\Omega \int_\Omega \frac{|f(x) - f(y)|^2}{|x - y|^{n+2s}} \, dx \, dy \geq 2\kappa_{n,2s} \int_\Omega \frac{|f(x)|^2}{M_{s,2}(x)^{2s}} \, dx, \tag{6.2.20}
$$

where $\kappa_{n,2s}$ is the sharp constant

$$
\kappa_{n,2s} := \frac{\pi^{(n-1)/2}\Gamma\left(\frac{1+ps}{2}\right)}{\Gamma\left(\frac{n+ps}{2}\right)} \frac{1}{s}\left\{ \frac{2^{-2s}}{\sqrt{\pi}} \Gamma\left(\frac{1+2s}{2}\right) \Gamma(1-s) - \frac{1}{2} \right\}. \tag{6.2.21}
$$

This refinement is achieved through the use of the one-dimensional inequality in [130], Theorem 2.1, to replace Lemma 6.9 above, that for $1/2 < s < 1$ and $f \in C_0^\infty(a, b)$,

$$\int_a^b \int_a^b \frac{|f(x) - f(y)|^2}{|x - y|^{1+2s}} \, dx \, dy \geq 2\kappa_{1,2s} \int_a^b |f(x)|^2 \left(\frac{1}{x - a} + \frac{1}{b - x} \right)^{2s} dx,$$
(6.2.22)

which has the corollary that for any open set $J \subset \mathbb{R}$, $1/2 < s < 1$ and $f \in C_0^\infty(J)$,

$$\int_J \int_J \frac{|f(x) - f(y)|^2}{|x - y|^{1+2s}} \, dx \, dy \geq 2\kappa_{1,2s} \int_J |f(x)|^2 \left(\frac{1}{\delta_J(x)} + \frac{1}{d_J(x)} \right)^{2s} dx, \quad (6.2.23)$$

where J is a countable union of disjoint intervals I_k and for $x \in J$, $d_J(x) = d_{I_k}(x) = \sup\{|t|: x + t \notin I_k\}$, where I_k is the unique interval containing x.

By the Sobolev inequality, the left-hand side of (6.2.2) dominates the L_q norm of f for $q = np/(n - ps)$. Dyda and Frank prove in [57] that this remains true even if the right-hand side of (6.2.2) is subtracted from the left; their result is the fractional Hardy–Sobolev–Maz'ya inequality in

Theorem 6.12 *Let $n \geq 2$, $2 \leq p < \infty$, $0 < s < 1$ and $1 < ps < n$. Then there exists a constant $k_{n,p,s} > 0$ such that for $q = np/(n - ps)$,*

$$\int_\Omega \int_\Omega \frac{|f(x) - f(y)|^p}{|x - y|^{n+ps}} \, dx \, dy - D_{n,p,s} \int_\Omega \frac{|f(x)|^p}{M_{s,p}(x)^{ps}} \, dx \geq k_{n,p,s} \left(\int_\Omega |f(x)|^q dx \right)^{p/q}$$
(6.2.24)

for all open $\Omega \subsetneq \mathbb{R}^n$ and all $f \in \overset{0}{W}_p^s(\Omega)$.

This is the fractional analogue of (5.5.6). In the case of $p = 2$ and $\Omega = \mathbb{R}_+^n$, a proof of (6.2.24) was given in [160]. A variant of (6.2.24) for a half-space, and more general John domains, is given in [59]; for $\mathbb{R}_+^n = \{x: x = (x_1, x_2, ..., x_n) \in \mathbb{R}^n, x_n > 0\}$, the integral on the right-hand side has a weight x_n^{-bq}, where $b = n(1/q - 1/p) + s$.

6.3 Fractional Hardy Inequality with a Remainder Term

Dyda [56] proved the following refinement of (6.2.20):

Theorem 6.13 *Let $1/2 < s < 1$ and Ω a bounded domain in $\mathbb{R}^n$. Then for all $u \in C_0^\infty(\Omega)$,*

$$\frac{1}{2} \int_\Omega \int_\Omega \frac{|f(x) - f(y)|^2}{|x - y|^{n+2s}} \, dx \, dy \geq \kappa_{n,2s} \int_\Omega \frac{|f(x)|^2}{M_{s,2}(x)^{2s}} \, dx$$

$$+ \frac{\lambda_{n,ps}}{\text{diam}\,(\Omega)} \int_\Omega \frac{|u(x)|^2}{M_{s-1/2}(x)^{2s-1}} \, dx,$$

$$(6.3.1)$$

where

$$\lambda_{n,ps} = \pi^{(n-1)/2} \Gamma \left(\frac{2s}{2} \right) \frac{4 - 2^{3-2s}}{2s\Gamma \left(\frac{n+2s-1}{2} \right)}. \qquad (6.3.2)$$

The constant $\kappa_{n,2s}$ cannot be replaced by a larger constant in (6.13).

The proof in [56] is based on the method developed in [130] to prove (6.2.20), but with the inequality in the following proposition used instead of (6.2.22).

Proposition 6.14 *Let $1 < \alpha < 2$ and $-\infty < a < b < \infty$. Then for all $u \in C_0^\infty (a, b)$,*

$$\frac{1}{2} \int_a^b \int_a^b \frac{(u(x) - u(y))^2}{|x - y|^{1+\alpha}} \, dx \, dy \geq \kappa_{1,\alpha} \int_a^b u(x)^2 \left(\frac{1}{x-a} + \frac{1}{b-x} \right)^{2s} dx$$

$$+ \frac{4 - 2^{3-\alpha}}{\alpha(b-a)} \int_a^b u(x)^2 \left(\frac{1}{x-a} + \frac{1}{b-x} \right)^{\alpha-1} dx.$$

$$(6.3.3)$$

The constant $\kappa_{1,\alpha}$ cannot be replaced by a larger one.

Proof An important first step is the calculation of

$$Lu(x) := \lim_{\varepsilon \to 0+} \int_{(-1,1) \cap \{y:\, |x-y|>\varepsilon\}} \frac{u(y) - u(x)}{|x - y|^{1+\alpha}} \, dy.$$

Let $q > -1$, $0 < \alpha < 2$ and $u_q(x) = (1 - x^2)^q$. Then on setting $t = y^2$ and integration by parts

$$Lu_q(0) = 2 \lim_{\varepsilon \to 0+} \int_\varepsilon^1 \frac{(1 - y^2)^q - 1}{y^{1+\alpha}} \, dy$$

$$= 2 \left(\frac{1}{2} \int_{\varepsilon^2}^1 (1 - t)^q t^{-1-\alpha/2} [(1 - t) + t] \, dt - \int_\varepsilon^1 y^{-1-\alpha} \, dy \right)$$

$$= 2 \lim_{\varepsilon \to 0+} \left(\frac{1}{\alpha} (1 - \varepsilon^2)^{q+1} \varepsilon^{-\alpha} - \frac{q+1}{\alpha} \int_{\varepsilon^2}^1 (1 - t)^q t^{-\alpha/2} \, dt \right)$$

$$+ 2 \lim_{\varepsilon \to 0+} \left(\frac{1}{2} \int_{\varepsilon^2} (1 - t)^q t^{-\alpha/2} \, dt + \frac{1}{\alpha} - \frac{\varepsilon^{-\alpha}}{\alpha} \right)$$

$$= \frac{2}{\alpha} [1 - (q + 1 - \alpha/2) B(q + 1, 1 - \alpha/2)]$$

since

$$\lim_{\varepsilon \to 0+} \left(\frac{1}{\alpha}(1-\varepsilon^2)^{q+1}\varepsilon^{-\alpha} - \frac{\varepsilon^{-\alpha}}{\alpha} \right) = \lim_{\varepsilon \to 0+} \frac{\varepsilon^{2-\alpha}}{\alpha} \left(\frac{(1-\varepsilon^2)^{q+1} - 1}{\varepsilon^2} \right) = 0;$$

B denotes the Euler beta function: $B(a,b) = \frac{\Gamma(a)\Gamma(b)}{\Gamma(1/2)}$.

For $x_0 \in (-1, 1)$,

$$Lu_q(x_0) = pv \int_{-1}^{1} \frac{(1-y^2)^q - (1-x_0^2)^q}{|y-x_0|^{1+\alpha}} \, dy$$

where pv denotes the principal value. The variable change

$$t = \varphi(y) = \frac{x_0 - y}{1 - x_0 y}$$

transforms Lu_p into

$$Lu_q(x_0) = (1-x_0^2)^{q-\alpha} \int_{-1}^{1} \frac{(1-t^2)^q - (1-tx_0)^{2q}}{|t|^{1+\alpha}}(1-tx_0)^{\alpha-1-2q} \, dt$$

$$= (1-x_0^2)^{q-\alpha} \left[Lu_q(0) - pv \int_{-1}^{1} \frac{(1-tx_0)^{\alpha-1} - 1}{|t|^{1+\alpha}} \, dt \right]$$

$$+ (1-x_0^2)^{q-\alpha} \left[pv \int_{-1}^{1} \frac{(1-tx_0)^{\alpha-1-2q} - 1}{|t|^{1+\alpha}}(1-t^2)^q \, dt \right].$$

Let

$$I := pv \int_{-1}^{1} \frac{(1-tx_0)^{\alpha-1} - 1}{|t|^{1+\alpha}} \, dt = \lim_{\varepsilon \to 0+} \left(J_\varepsilon(x_0) + J_\varepsilon(-x_0) \right),$$

where

$$J_\varepsilon(x_0) = \int_{\varepsilon}^{1} \frac{(1-tx_0)^{\alpha-1} - 1}{|t|^{1+\alpha}} \, dt = \int_{\varepsilon}^{1} \left(\frac{1}{t} - x_0 \right)^{\alpha-1} \frac{dt}{t^2} = \frac{\varepsilon^{-\alpha} - 1}{\alpha}$$

$$= \frac{1}{\alpha} \left(\frac{1}{\varepsilon} - x_0 \right)^{\alpha} - \frac{1}{\alpha}(1-x_0)^{\alpha} - \frac{\varepsilon^{-\alpha} - 1}{\alpha}$$

$$= \frac{1}{\alpha} - \frac{1}{\alpha}(1-x_0)^{\alpha} + \frac{(1-\varepsilon x_0)^{\alpha} - 1}{\alpha \varepsilon^{\alpha}}.$$

By l'Hôpital's rule

$$I = \frac{2}{\alpha} - \frac{1}{\alpha}(1-x_0)^{\alpha} - \frac{1}{\alpha}(1+x_0)^{\alpha}$$

and so

$$Lu_q(x) = \frac{(1-x^2)^{q-\alpha}}{\alpha} \{(1-x)^{\alpha} + (1+x^{\alpha})\}$$

$$- \frac{(1-x^2)^{q-\alpha}}{\alpha} \{(2q+2-\alpha)B(q+1, 1-\alpha/2) + \alpha I(q)\}, \quad (6.3.4)$$

where

$$I(q) := pv \int_{-1}^1 \frac{(1 - tx_0)^{\alpha-1-2q} - 1}{|t|^{1+\alpha}} (1 - t^2)^q dt.$$

We also have

$$I(\alpha/2) = pv \int_{-1}^1 \frac{(1 - tx)^{-1} - 1}{|t|^{1+\alpha}} (1 - t^2)^{\alpha/2} dt$$

$$= \int_{-1}^1 \frac{\sum_{k=2}^\infty (tx)^k}{|t|^{1+\alpha}} (1 - t^2)^{\alpha/2} dt$$

$$= 2 \int_{-1}^1 \frac{\sum_{k=2}^\infty (tx)^k}{|t|^{1+\alpha}} (1 - t^2)^{\alpha/2} dt$$

$$= \sum_{k=1}^\infty B(k - \alpha/2, 1 + \alpha/2) x^{2k}$$

$$= \Gamma(1+\alpha/2)\Gamma(-\alpha/2) \left(\sum_{k=0}^\infty \frac{x^{2k}\Gamma(k-\alpha/2)}{\Gamma(-\alpha/2)k!} - 1 \right)$$

$$= \frac{2B(1+\alpha/2, 1-\alpha/2)}{\alpha} (1 - (1 - x^2)^{\alpha/2}).$$

It can also be shown that $I(\frac{\alpha-1}{2}) = I(\frac{\alpha-2}{2}) = 0$ and, if $1 < \alpha < 2$, that $I(\frac{\alpha-3}{2}) = x^2 B(\frac{\alpha-1}{2}, 1 - \frac{\alpha}{2})$.

The next step in the proof of the proposition is the application of a result which is analogous to the ground state representation for half-spaces and $\mathbb{R}^n \setminus \{0\}$ in [82] and [83], and may be considered as a special case of Proposition 2.3 in [82]. The result is that with $0 < \alpha < 2$, $w(x) = (1 - x^2)^{(\alpha-1)/2}$ and $u \in C_0((-1, 1))$,

$$\frac{1}{2} \int_{-1}^1 \int_{-1}^1 \frac{(u(x) - u(y))^2}{|x - y|^{1+\alpha}} \, dx \, dy$$

$$= \frac{1}{2} \int_{-1}^1 \int_{-1}^1 \left(\frac{u(x)}{w(x)} - \frac{u(y)}{w(y)} \right)^2 \frac{w(x)w(y)}{|x - y|^{1+\alpha}} \, dx \, dy$$

$$+ 2^\alpha \kappa_{1,\alpha} \int_{-1}^1 u(x)^2 (1 - x^2)^{-\alpha} \, dx$$

$$+ \frac{1}{\alpha} \int_{-1}^1 u(x)^2 \left[2^\alpha - (1 + x)^\alpha - (1 - x)^\alpha \right] (1 - x^2)^{-\alpha} \, dx. \quad (6.3.5)$$

We are now equipped to complete the proof of Proposition 6.14. By scaling we may and shall assume that $a = -1$, $b = 1$. By (6.3.5), we require that

$$2^\alpha - (1 + x)^\alpha - (1 - x)^\alpha \geq (2^\alpha - 2)(1 - x^2), \quad 1 \leq \alpha \leq 2, \ 0 \leq x \leq 1. \quad (6.3.6)$$

On substituting $u = x^2$, it suffices to prove that

$$g(u) := (2^\alpha - 2)u - (1 - \sqrt{u})^\alpha - (1 + \sqrt{u})^\alpha + 2$$

is concave, or

$$g'(u) = 2^\alpha - 2 + \frac{\alpha}{2\sqrt{u}} \left((1 - \sqrt{u})^{\alpha-1} - (1 + \sqrt{u})^{\alpha-1} \right)$$

is decreasing. Setting $u = t^2$, $h(t) = (1 - t)^{\alpha-1} - (1 + t)^{\alpha-1}$, we have that

$$\frac{(1 - t)^{\alpha-1} - (1 + t)^{\alpha-1}}{t} = \frac{h(t) - h(0)}{t}.$$

Since h is concave, the function $t \mapsto \frac{h(t)-h(0)}{t}$ is decreasing and hence so is g'. Therefore (6.3.6) is proved and the proposition follows. The sharpness of $\kappa_{1,\alpha}$ is already established in [130]. $\square$

Proof of Theorem 6.13 This follows by using Proposition 6.14 with $\alpha = 2s$ instead of (6.2.22) in the proof of the case $p = 2$ of Theorem 6.8.

7

Classical and Fractional Inequalities of Rellich Type

7.1 The Classical Inequalities

Rellich's classical inequality in [153] asserts that for all $u \in C_0^\infty(\mathbb{R}^n \setminus \{0\})$ and $n \in \mathbb{N} \setminus \{2\}$,

$$\int_{\mathbb{R}^n} |\Delta u(x)|^2 \, dx \geq \frac{n^2(n-4)^2}{16} \int_{\mathbb{R}^n} \frac{|u(x)|^2}{|x|^4} \, dx, \tag{7.1.1}$$

with sharp constant $n^2(n-4)^2/16$. The inequality also holds for $n = 2$ (with constant 1), but only for those functions $u \in C_0^\infty(\mathbb{R}^2 \setminus \{0\})$ which, in terms of polar co-ordinates (r, θ), satisfy

$$\int_0^\infty u(r, \theta) \cos \theta \, d\theta = \int_0^\infty u(r, \theta) \sin \theta \, d\theta = 0. \tag{7.1.2}$$

In [73], Rellich-type inequalities involving magnetic Laplacians with magnetic potentials of Aharonov–Bohm type are studied, which are valid for all $n \in \mathbb{N}$ in some circumstances. What is of particular significance for (7.1.1) is that they clarify the situation for the case $n = 2$ and also the trivial case $n = 4$. The study was motivated by the Laptev–Weidl inequality (5.1.9) in which a magnetic Hardy inequality is shown to be valid in the case $n = 2$ (when there is no nontrivial Hardy inequality) if the magnetic potential is of Aharonov–Bohm type

$$\mathbf{A}(x) = \Psi \left(-\frac{x_2}{|x|^2}, \frac{x_1}{|x|^2} \right), x = (x_1, x_2) \tag{7.1.3}$$

with non-integer flux Ψ; the magnetic field curl $\mathbf{A} = 0$ in $\mathbb{R}^2 \setminus \{0\}$. The following two theorems are proved in [73]: in them

$$\Delta_\mathbf{A} = (\nabla - \mathbf{A})^2 \text{ is the magnetic Laplacian.}$$

Theorem 7.1 *For all* $u \in C_0^\infty(\mathbb{R}^2 \setminus \{0\})$,

$$\int_{\mathbb{R}^2} |\Delta_\mathbf{A} u(x)|^2 \frac{dx}{|x|^s} \, dx \geq C(2, s) \int_{\mathbb{R}^2} |u(x)|^2 \frac{dx}{|x|^{s+4}} \, dx, \tag{7.1.4}$$

where

$$C(2, s) = \min_{m \in \mathbb{Z}} \left\{ (m + \Psi)^2 - \frac{(s+2)^2}{4} \right\}^2 . \tag{7.1.5}$$

If $\Psi \notin \mathbb{Z}$ ($\Psi \in (0, 1)$ without loss of generality), we have

$$C(2, 0) = \min\{(\Psi^2 - 1)^2, \Psi^2(\Psi - 2)^2\}$$

$$= \begin{cases} (\Psi^2 - 1)^2 & \text{if } \Psi \in [\frac{1}{2}, 1), \\ \Psi^2(\Psi - 2)^2 & \text{if } \Psi \in [0, \frac{1}{2}). \end{cases} \tag{7.1.6}$$

Remark 7.2

If $\Psi \in \mathbb{Z}$, then $C(2, 0) = 0$. However, if (7.1.2) is satisfied, then the minimum in (7.1.5) is over $m \in \mathbb{Z} \setminus \{-1, 1\}$ and this recovers the result $C(2, 0) = 1$.

Theorem 7.3　*Let $u \in C_0^\infty(\mathbb{R}^4 \setminus \mathcal{L}_4)$, where $\mathcal{L}_4 := \{x = (r, \theta_1, \theta_2, \theta_3; \theta_1, \theta_2 \in (0, \pi), \theta_3 \in (0, 2\pi) : r \sin\theta_1 \sin\theta_2 = 0\}$. Then $\text{curl } A = 0$ on $\mathbb{R}^4 \setminus \mathcal{L}_4$ and*

$$\int_{\mathbb{R}^4} |\Delta_A u(x)|^2 \frac{dx}{|x|^s} \geq C(4, s) \int_{\mathbb{R}^4} |u(x)|^2 \frac{dx}{|x|^{s+4}}, \tag{7.1.7}$$

where

$$C(4, s) := \inf_{m \in \mathbb{Z}'} \left\{ \left[(m - \Psi)^2 - 1 - \frac{s(s+4)}{4} \right]^2 \right\}, \tag{7.1.8}$$

and $\mathbb{Z}' := \{m \in \mathbb{Z} : (m - \Psi)^2 \geq 1\}$. In particular, when $s = 0$ and $\Psi \in (0, 1)$,

$$C(4, 0) = \min\{[(1 - \Psi)^2 - 1]^2, [(-2 - \Psi)^2 - 1]^2\} > 0.$$

When $\Psi = 0$, (7.1.7) is satisfied on $C_0^\infty(\mathbb{R}^4 \setminus \{0\})$. The inequality is trivial if $C(4, 0) = 0$, but there is a restricted class of functions which is such that the infimum is attained for $m = \pm 2$, and so $C(4, 0) = 9$; this is an analogue for $n = 4$ of the result for $n = 2$ in Remark 7.2. We refer to [15], Corollary 6.4.10, for further details.

Similar results to Theorems 7.1 and 7.3 are given in the case $n = 3$ in [15], and for $n > 4$ in [166].

Next, we consider Rellich inequalities on a domain $\Omega \subset \mathbb{R}^n$, $n \geq 2$. Let Ω be a proper, non-empty open subset of $\mathbb{R}^n (n \geq 2)$ and $\delta(x) := \text{dist}(x, \partial\Omega)$, the distance from $x \in \Omega$ to the boundary of Ω. The following Rellich inequality in $L_2(\Omega)$ is established in [15], Corollary 6.2.7,

$$\int_\Omega |\Delta u(x)|^2 dx \geq \frac{9}{16} \int_\Omega \frac{|u(x)|^2}{\delta(x)^4} dx, \quad u \in C_0^\infty(\Omega), \tag{7.1.9}$$

under the assumption that δ is superharmonic, i.e., $\Delta \delta \leq 0$ in the distributional sense; this requirement is met if Ω is convex or if Ω is weakly mean convex with

$\Sigma(\Omega) = \Omega \setminus G(\Omega)$ a null set; see Section 5.3. The proof in [15] is based on the abstract Hardy-type inequality

$$\int_\Omega |\Delta V(x)||u(x)|^2 dx \leq 4 \int_\Omega \frac{|\nabla V(x)|^2}{|\Delta V(x)|} |\nabla u(x)|^2 dx, \quad u \in C_0^\infty(\Omega),$$

which was proved for ΔV of one sign in [123], Lemma 2 (see also [49]). For $s \neq 0$, choose $V(x) = -[(s+1)/s]\delta(x)^{-s}$ and for $s = 0$ let $V(x) = \ln \delta(x)$. Then $|\nabla V(x)|^2 = (s+1)^2 \delta(x)^{-2(s+1)}$, and when $\Delta \delta(x) \leq 0$,

$$-\Delta V(x) = (s+1)^2 \delta(x)^{-(s+2)} + (s+1)\delta(x)^{-(s+1)}(-\Delta\delta(x)) \geq (s+1)^2 \delta(x)^{-(s+2)}.$$

It follows that for $n \geq 2$,

$$(s+1)^2 \int_\Omega \frac{|u(x)|^2}{\delta(x)^{s+2}} dx \leq 4 \int_\Omega \frac{|\nabla u(x)|^2}{\delta(x)^s} ds, \quad u \in C_0^\infty(\Omega). \tag{7.1.10}$$

We show that (7.1.9) is a consequence of (7.1.10). With the notation $u_j = \partial_j u$, $u_{jk} = \partial_j \partial_k u$ and $u_{jkl} = \partial_j \partial_k \partial_l u$, we have

$$\begin{aligned}
\int_\Omega |\Delta u(x)|^2 dx &= \Sigma_{j,k=1}^n \int_\Omega u_{jj} \overline{u_{kk}} \, dx \\
&= \Sigma_{j=k} \int_\Omega |u_{jj}|^2 dx - \Sigma_{j \neq k} \int_\Omega u_j \overline{u_{jkk}} \, dx \\
&= \Sigma_{j=k} \int_\Omega |u_{jj}|^2 dx + \Sigma_{j \neq k} \int_\Omega u_{jk} \overline{u_{jk}} \, dx \\
&= \Sigma_{j=1}^n \int_\Omega |\nabla(u_j)|^2 \, dx. \tag{7.1.11}
\end{aligned}$$

Hence from (7.1.10),

$$\begin{aligned}
\int_\Omega |\Delta u(x)|^2 \, dx &= \Sigma_{j=1}^n \frac{1}{4} \int_\Omega \frac{|\nabla u(x)|^2}{\delta(x)^2} \, dx \\
&\geq \frac{9}{16} \int_\Omega \frac{|u(x)|^2}{\delta(x)^4} \, dx,
\end{aligned}$$

thus (7.1.9),

In the L_p setting, results have been obtained by Davies and Hinz in [49]. Their main tool is an abstract Rellich-type inequality reminiscent of the abstract Hardy-type inequality in [123], but now depending on the existence of a positive function V which is such that $\Delta V < 0$ and $\Delta(V^a) \leq 0$ for some $a > 1$: this is that if $p \in (1, \infty)$ and $u \in C_0^\infty(\Omega)$,

$$\int_\Omega |\Delta V(x)||u(x)|^p \, dx \leq \left(\frac{p^2}{(p-1)a+1}\right)^p \int_\Omega \frac{V^p(x)}{|\Delta V(x)|^{p-1}} |\Delta u(x)|^p \, dx.$$

When $\Omega = \mathbb{R}^n \setminus \{0\}$ and $n > s > 2$, the choice $V(x) = |x|^{-(s-2)}$, $a = \frac{n-2}{s-2}$ gives

$$\int_{\mathbb{R}^n} \frac{|\Delta u(x)|^p}{|x|^{s-2p}} \, dx \geq \left(p^{-2}(n-s)[(p-1)n+s-2p]\right)^p \int_{\mathbb{R}^n} \frac{|u(x)|^p}{|x|^s} \, dx,$$

whence the Rellich inequality

$$\int_{\mathbb{R}^n} |\Delta u(x)|^p dx \geq \left(\frac{n(p-1)(n-2p)}{p^2}\right)^p \int_{\mathbb{R}^n} \frac{|u(x)|^p}{|x|^{2p}} \, dx; \qquad (7.1.12)$$

the constant is shown to be sharp in [15], Corollary 6.3.5.

In [63], the mean distance function M_p defined in (6.2) by

$$\frac{1}{M_p(x)^p} := \frac{\sqrt{\pi}\,\Gamma\left(\frac{n+p}{2}\right)}{\Gamma\left(\frac{p+1}{2}\right)\Gamma\left(\frac{n}{2}\right)} \int_{\mathbb{S}^{n-1}} \frac{1}{\delta_v^p(x)} \, d\omega(v) \qquad (7.1.13)$$

is used to obtain Rellich inequalities of the form

$$\int_\Omega |\Delta u(x)|^p \, dx \geq C \int_\Omega \frac{|u(x)|^p}{M_{2p}(x)^{2p}} \, dx, \quad u \in C_0^\infty(\Omega),$$

for general domains Ω. The following is proved in [63]:

Theorem 7.4 *Let Ω be a non-empty, proper, open subset of $\mathbb{R}^n$, let $p \in (1, \infty)$ and suppose that $u \in C_0^2(\Omega)$. Then*

$$\int_\Omega \frac{|u(x)|^p}{M_{2p}(x)^{2p}} \, dx \leq K(p,n) \int_\Omega |\Delta u(x)|^p \, dx, \qquad (7.1.14)$$

where

$$K(p,n) = c_p B(n, 2p) n^d \cot^{2p}\left(\frac{\pi}{2p^*}\right). \qquad (7.1.15)$$

Here

$$d = 2 \text{ if } 1 < p < 2, \ d = 2p/p' \text{ if } 2 < p < \infty,$$

$$p^* = \max\{p, p'\}, c_p = \left(\frac{p}{2p-1}\right)^p \left(\frac{p}{p-1}\right)^p,$$

and

$$B(n, 2p) = \frac{\sqrt{\pi}\,\Gamma\left(\frac{n+2p}{2}\right)}{\Gamma\left(\frac{2p+1}{2}\right)\Gamma\left(\frac{n}{2}\right)}.$$

If $p = 2$, then

$$\int_\Omega \frac{|u(x)|^2}{M_4(x)^4} \, dx \leq \frac{16}{9} \int_\Omega |\Delta u(x)|^2 \, dx. \qquad (7.1.16)$$

For Ω convex, $M_{2p}(x) \leq \delta(x) := \inf\{|y-x| : y \in \mathbb{R}^n \setminus \Omega\}$.

We refer to [63] for a proof but some comments might be helpful. The cotangent factor in (7.1.15) appears in the inequality

$$\|D_j D_k u\|_p \leq \cot^2\left(\frac{\pi}{2p*}\right)\|\Delta u\|_p, \ u \in C_0^\infty(\mathbb{R}^n)$$

for $j, k = 1, 2, ..., n$, which follows from the identity

$$D_j D_k u = -R_j R_k \Delta u$$

involving the Riesz transform R_j in $L_p(\mathbb{R}^n)$ and the remarkable result

$$\|R_j : L_p(\mathbb{R}^n) \to L_p(\mathbb{R}^n)\| = \cot^2\left(\frac{\pi}{2p*}\right)$$

proved in [105] and [16]. A brief background discussion of Riesz transforms may be found in [65], Section 1.4.

7.2 Fractional Rellich Inequalities in $\mathbb{R}^n$

Frank and Seiringer establish the following Hardy inequality in [82], Theorem 1.1. Let $0 < s < 1$ and suppose that $u \in C_0^\infty(\mathbb{R}^n)$ if $1 \leq p < n/s$, while $u \in C_0^\infty(\mathbb{R}^n\backslash\{0\})$ if $n/s < p < \infty$. Then

$$\int_{\mathbb{R}^n}\int_{\mathbb{R}^n}\frac{|u(x)-u(y)|^p}{|x-y|^{n+ps}}\,dx\,dy \geq C_{n,s,p}\int_{\mathbb{R}^n}\frac{|u(x)|^p}{|x|^{ps}}\,dx, \tag{7.2.1}$$

where

$$C_{n,s,p} := 2\int_0^1 r^{ps-1}\left|1-r^{(n-ps)/p}\right|^p \Phi_{n,s,p}(r)\,dr, \tag{7.2.2}$$

$$\Phi_{n,s,p}(r) = |\mathbb{S}^{n-2}|\int_{-1}^1\frac{(1-t^2)^{(n-3)/2}}{(1-2rt+r^2)^{(n+ps)/2}}\,dt \text{ if } n \geq 2, \tag{7.2.3}$$

and

$$\Phi_{1,s,p}(r) = (1-r)^{-1-ps}+(1+r)^{-1-ps} \text{ if } n = 1. \tag{7.2.4}$$

In the case $p = 2$, (7.2.1) follows from Proposition 3.28, and the identity

$$\left\|(-\Delta)^{s/2}u\right\|_{2,\mathbb{R}^n}^2 = C(n,s)\int_{\mathbb{R}^n}\int_{\mathbb{R}^n}\frac{|u(x)-u(y)|^2}{|x-y|^{n+2s}}\,dx\,dy, \tag{7.2.5}$$

in which

$$C(s,n) = 2^{2s}\pi^{-n/2}\Gamma\left(\frac{n}{2}+s\right)/|\Gamma(-s)| \tag{7.2.6}$$

and thus

$$\lim_{s\to 1-} C(s,n)/(1-s) = 2n\pi^{-n/2}\Gamma(n/2) = 4n/\omega_{n-1}.$$

For $s \in (0,1)$, $p = 2$, $n > 2s$ and $u \in C_0^\infty(\mathbb{R}^n)$ (7.2.1) is then a consequence of (7.2.5) and the Herbst inequality [96]

$$\int_{\mathbb{R}^n} \left|(-\Delta)^{s/2} u(x)\right|^2 dx \geq C_{s,n} \int_{\mathbb{R}^n} \frac{|u(x)|^2}{|x|^{2s}} dx, \tag{7.2.7}$$

which has the sharp constant

$$C_{s,n} = 2^{2s} \frac{\Gamma^2\left(\frac{n+2s}{4}\right)}{\Gamma^2\left(\frac{n-2s}{4}\right)}. \tag{7.2.8}$$

Therefore, for $s \in (0,1)$, $n > 2s$ and $u \in C_0^\infty(\mathbb{R}^n)$,

$$\int_{\mathbb{R}^n}\int_{\mathbb{R}^n} \frac{|u(x)-u(y)|^2}{|x-y|^{n+2s}} dx\,dy \geq C_{n,s,2} \int_{\mathbb{R}^n} \frac{|u(x)|^2}{|x|^{2s}} dx, \tag{7.2.9}$$

with sharp constant

$$C_{n,s,2} = 2C_{s,n}/C(s,n) = 2\pi^{n/2}\frac{\Gamma^2\left(\frac{n+2s}{4}\right)|\Gamma(-s)|}{\Gamma^2\left(\frac{n-2s}{4}\right)\Gamma\left(\frac{n+2s}{4}\right)}. \tag{7.2.10}$$

Another consequence of (7.2.5) is

Corollary 7.5 *Let $\sigma \in (0,1)$ and $n > 2\sigma$. Then for all $C_0^\infty(\mathbb{R}^n)$,*

$$\sum_{i=1}^{n}\int_{\mathbb{R}^n}\int_{\mathbb{R}^n} \frac{|D_i u(x)-D_i u(y)|^2}{|x-y|^{n+2\sigma}} dx\,dy \geq 2C(\sigma,n)^{-1}C_{\sigma+1,n}\int_{\mathbb{R}^n}\frac{|u(x)|^2}{|x|^{2+2\sigma}}dx, \tag{7.2.11}$$

where the constant is sharp.

Proof From (7.2.5) with $s = 1 + \sigma$ and $\sigma \in (0,1)$, we have

$$\sum_{i=1}^{n}\int_{\mathbb{R}^n}\int_{\mathbb{R}^n} \frac{|D_i u(x)-D_i u(y)|^2}{|x-y|^{n+2\sigma}} dx\,dy$$

$$= 2C(\sigma,n)^{-1}\sum_{i=1}^{n}\int_{\mathbb{R}^n} |\xi|^{2\sigma}|(F(D_i u))(\xi)|^2 d\xi$$

$$= 2C(\sigma,n)^{-1}\sum_{i=1}^{n}\int_{\mathbb{R}^n} \left|F^{-1}\left(|\xi|^{\sigma+1}F(u)(\xi)\right)\right|^2 d\xi$$

$$= 2C(\sigma,n)^{-1}\left\|(-\Delta)^{(\sigma+1)/2}u\right\|_{2,\mathbb{R}^n}^2$$

$$\geq 2C(\sigma,n)^{-1}C_{\sigma+1,n}\int_{\mathbb{R}^n}\frac{|u(x)|^2}{|x|^{2+2\sigma}}dx. \tag{7.2.12}$$

$\square$

Note that we also have from (7.2.1)

$$\sum_{i=1}^{n} \int_{\mathbb{R}^n} \int_{\mathbb{R}^n} \frac{|D_i u(x) - D_i u(y)|^2}{|x-y|^{n+2\sigma}} \, dx \, dy$$

$$\geq C_{n,\sigma,2} \int_{\mathbb{R}^n} \sum_{i=1}^{n} \frac{|D_i u(x)|^2}{|x|^{2\sigma}} \, dx$$

$$= C_{n,\sigma,2} \int_{\mathbb{R}^n} \frac{|\nabla u(x)|^2}{|x|^{2\sigma}} \, dx$$

$$\geq C_{n,\sigma,2} \left| \frac{2+2\sigma-n}{2} \right|^2 \int_{\mathbb{R}^n} \frac{|u(x)|^2}{|x|^{2+2\sigma}} \, dx,$$

the final step being the weighted Hardy inequality (7.1.10). Since the constant in (7.2.12) is sharp, it follows that

$$C_{n,\sigma,2} \left| \frac{2+2\sigma-n}{2} \right|^2 \leq 2C(\sigma,n)^{-1} C_{\sigma+1,n};$$

hence by (7.2.10),

$$C_{\sigma,n} \left| \frac{2+2\sigma-n}{2} \right|^2 \leq C_{\sigma+1,n}. \tag{7.2.13}$$

For $n > 2$ the inequality (7.2.13) is strict since, on allowing $\sigma \to 1-$, the left-hand side tends to $((n-2)(n-4))^2/16$ while the right-hand side tends to $(n(n-4))^2/16$, the optimal constant in the Rellich inequality (7.1.1). If $\sigma \to 0+$, (7.2.11) becomes the Hardy inequality and the constants on both sides of (7.2.13) tend to the optimal Hardy constant $(n-2)^2/4$.

Remark 7.6

The inequality (7.2.5) is the special case $p = 2$ of Herbst's inequality in [96] which is, for $1 < p < \infty$, $s > 0$, $n > ps$ and $u \in C_0^\infty (\mathbb{R}^n)$, that

$$\int_{\mathbb{R}^n} \frac{|u(x)|^p}{|x|^{ps}} \, dx \leq K_{n,p,s}^p \left\| (-\Delta)^{s/2} u \right\|_{p,\mathbb{R}^n}^p, \tag{7.2.14}$$

with best possible constant

$$K_{n,p,s} = 2^{-s} \frac{\Gamma\left(\frac{n(p-1)}{2p}\right) \Gamma\left(\frac{n-ps}{2p}\right)}{\Gamma\left(\frac{n}{2p}\right) \Gamma\left(\frac{n(p-1)+ps}{2p}\right)}. \tag{7.2.15}$$

This is also established in [156]; moreover, Samko determines a sharp constant for the Hardy–Stein–Weiss inequality for fractional Riesz operators in $L_p (\mathbb{R}^n, \rho)$ with a power weight $\rho(x) = |x|^\beta$ and as a corollary finds the sharp constant for a similar weighted inequality for fractional powers of the Laplace–Beltrami operator on the unit sphere. A proof of (7.2.14) in the case $p = 2$ was given in [172]; moreover, Yafaev shows that if $n < 2(1 + \sigma), 1 + \sigma - n/2 \notin \mathbb{Z}$ and $k := \left[1 + \sigma - n/2\right]$, then

$$
\int_{\mathbb{R}^n} |x|^{-2-2\sigma} \left| u(x) - \sum_{|\alpha| \leq k} (\alpha!)^{-1} \left(D^\alpha u\right)(0) x^\alpha \right|^2 dx
$$

$$
\leq K_{n,\sigma}^2 \left\| (-\Delta)^{(1+\sigma)/2} u \right\|_{2,\mathbb{R}^n}^2,
$$

where

$$
K_{n,\sigma} = 2^{-1-\sigma} \max \left\{ \frac{\Gamma\left(\frac{n-2-2\sigma}{4}\right)}{\Gamma\left(\frac{n+2+2\sigma}{4}\right)}, \frac{\Gamma\left(\frac{n-2\sigma}{4}\right)}{\Gamma\left(\frac{n+4+2\sigma}{4}\right)} \right\}.
$$

Thus in particular, (7.2.14), with $p = 2$, $s = 1 + \sigma$ and constant $K_{n,\sigma}^2$, holds for all $u \in C_0^\infty (\mathbb{R}^n \backslash \{0\})$ if $n < 2(1 + \sigma)$ and $1 + \sigma - n/2 \notin \mathbb{Z}$.

7.3 Fractional Rellich Inequalities in General Domains

The mean distance defined in (6.2.1), namely

$$
\frac{1}{M_{s,p}(x)^{ps}} := \frac{\pi^{1/2} \Gamma\left(\frac{n+ps}{2}\right)}{\Gamma\left(\frac{1+ps}{2}\right) \Gamma\left(\frac{n}{2}\right)} \int_{\mathbb{S}^{n-1}} \frac{1}{\delta_v^{ps}(x)} d\omega(v), \tag{7.3.1}
$$

will again feature prominently in this and following sections, with the range of the parameter s specific to the problem being considered. It remains true that if Ω is convex with non-empty boundary, then $M_{s,p}(x) \leq \delta(x) := \inf\{|y - x| : y \notin \Omega\}$.

The main theorem on fractional Rellich inequalities comes from [66] and sets the scene for much of what follows in this chapter. The method of proof is that in [130] which was what was used to prove Theorem 6.8 and (6.2.20). We shall use the following notation, some of which is reminiscent of that in Section 6.2. Suppose that $1 < p < \infty, 1/p < \sigma < 1$ and define

$$
e_p(n) := \begin{cases} 1 & \text{if } p \leq 2, \\ n^{-(p-2)/2} & \text{if } p > 2, \end{cases} \tag{7.3.2}
$$

and

$$
\mathcal{D}_{n,p,\alpha} := \frac{\pi^{(n-1)/2} \Gamma\left(\frac{1+p+\alpha}{2}\right)}{\Gamma\left(\frac{n+p+\alpha}{2}\right)} \mathcal{D}_{1,p,\alpha}, \tag{7.3.3}
$$

where

$$\mathcal{D}_{1,p,\alpha} = 2 \int_0^1 \frac{\left|1 - r^{(\alpha-1)/p}\right|^p}{(1-r)^{1+\alpha}} \, dr. \tag{7.3.4}$$

In the special case in which $p = 2$, we have from the appendix of [22] that

$$\mathcal{D}_{1,2,\alpha} = \frac{2}{\alpha} \left\{ \frac{2^{-\alpha}}{\sqrt{\pi}} \Gamma\left(\frac{1+\alpha}{2}\right) \Gamma\left(\frac{2-\alpha}{2}\right) - 1 \right\}. \tag{7.3.5}$$

Theorem 7.7 *Let Ω be an open subset of $\mathbb{R}^n$ with non-empty boundary, let $p \in (1, \infty)$ and suppose that $\sigma \in (1/p, 1)$. Then for all $f \in C_0^\infty(\Omega)$,*

$$\sum_{i=1}^n \int_\Omega \int_\Omega \frac{|D_i f(x) - D_i f(y)|^p}{|x-y|^{n+p\sigma}} \, dx \, dy$$

$$\geq e_p(n) \mathcal{D}_{n,p,p\sigma} \left(\sigma + 1 - 1/p\right)^p \int_\Omega \frac{|f(x)|^p}{M_{1+\sigma,p}(x)^{p+p\sigma}} \, dx. \tag{7.3.6}$$

The following improvement is possible in the case $p = 2$ with $\sigma \in (1/2, 1)$:

$$\sum_{i=1}^n \int_\Omega \int_\Omega \frac{|D_i f(x) - D_i f(y)|^2}{|x-y|^{n+2\sigma}} \, dx \, dy$$

$$\geq 2\kappa_{n,2\sigma} \left(\sigma + 1/2\right)^2 \int_\Omega \frac{|f(x)|^2}{M_{1+\sigma,2}(x)^{2+2\sigma}} \, dx, \tag{7.3.7}$$

where

$$2\kappa_{n,2\sigma} := 2 \frac{\pi^{(n-1)/2} \Gamma\left(\frac{3+2\sigma}{2}\right)}{\sigma \Gamma\left(\frac{n+2+2\sigma}{2}\right)} \left\{ \frac{2^{-2\sigma}}{\sqrt{\pi}} \Gamma\left(\frac{1+2\sigma}{2}\right) \Gamma\left(\frac{2-2\sigma}{2}\right) - \frac{1}{2} \right\}$$

$$> \mathcal{D}_{n,2,2\sigma}. \tag{7.3.8}$$

The proof of this theorem will be given later, after the establishment of various one-dimensional inequalities based on ones in [130]. Only the case $k = 1$ of the first lemma is needed in this section, but the general case will be required when we consider higher-order inequalities later.

Lemma 7.8 *Let $-\infty < a < b < \infty$, $p + s - 1 > 0$, $1 < p < \infty$, $k \in \mathbb{N}$ and for $t \in (a, b)$, set $\delta_{(a,b)}(t) := \min\{t - a, b - t\}$. Then, for all $f \in C_0^\infty(a, b)$,*

$$\int_a^b |f^k(x)|^p \left\{ \frac{1}{(x-a)} + \frac{1}{(b-x)} \right\}^s$$

$$\geq \Pi_{j=1}^k \left(\frac{jp + s - 1}{p} \right)^p \int_a^b \frac{|f(x)|^p}{\delta_{(a,b)}(x)^{kp+s}} \, dx. \tag{7.3.9}$$

Proof   Let $c = (a+b)/2$, $j \in \{1, 2, \cdots, k\}$ and $q = p/(p-1)$. On integration by parts and the use of Hölder's inequality, we have (cf. [66], Lemma 4.7)

$$
\begin{aligned}
I_a &:= \int_a^c \frac{|f(x)|^p}{|x-a|^{jp+s}}\,dx \\
&= \frac{p}{jp+s-1} \int_a^c |f(x)|^{p-2}\Re[\bar{f}f']\left[(x-a)^{-(jp+s)+1} - (c-a)^{-(jp+s)+1}\right] \\
&\le \frac{p}{jp+s-1} \int_a^c \frac{|f|^{p-1}}{(x-a)^{(jp+s)/q}}(x-a)^{(jp+s)/q}|f'|[(x-a)^{-(jp+s)+1} \\
&\quad - (c-a)^{-(jp+s)+1}]\,dx \\
&\le \frac{p}{jp+s-1} \int_a^c \left(\frac{|f|^p}{(x-a)^{jp+s}}\right)^{1/q} |f'|(x-a)^{-(jp+s)/p+1} \\
&\quad \times \left[1 - \left(\frac{x-a}{c-x}\right)^{jp+s-1}\right]dx.
\end{aligned}
$$

Therefore,

$$
\begin{aligned}
I_a \le &= \left(\frac{p}{jp+s-1}\right)^p \int_a^c \frac{|f'|^p}{(x-a)^{(j-1)p+s}}\left[1 - \left(\frac{x-a}{c-a}\right)^{jp+s-1}\right]^p dx \\
&\le \left(\frac{p}{jp+s-1}\right)^p \int_a^c \frac{|f'|^p}{(x-a)^{(j-1)p+s}}\,dx. \qquad (7.3.10)
\end{aligned}
$$

We also have for $j = 1$,

$$
I_a \le \left(\frac{p}{p+s-1}\right)^p \int_a^c \frac{|f'|^p}{(x-a)^s}\,dx\left[1 - \left(\frac{x-a}{c-x}\right)^{p+s-1}\right]^p dx.
$$

Let

$$
h_a(x) := \left[1 - \left(\frac{x-a}{c-a}\right)^{p+s-1}\right]^p - \left[1 + \left(\frac{x-a}{b-x}\right)\right]^s.
$$

Then, $h_a(a) = 0$, $h_a(c) = -2^s$ and $h_a'(x) \le 0$ on (a, c), and so $h_a(x) < 0$ on (a, c). Hence when $j = 1$,

$$
\begin{aligned}
I_a &\le \int_a^c \frac{|f'(x)|^p}{(x-a)^s}\left\{h_a(x) + \left[1 + \left(\frac{x-a}{b-x}\right)\right]^s\right\}dx \\
&= \int_a^c |f'(x)|^p \left\{\frac{1}{(x-a)} + \frac{1}{(b-x)}\right\}^s dx. \qquad (7.3.11)
\end{aligned}
$$

Similar inequalities to (7.3.10) and (7.3.11) hold for the integral

$$
I_b := \int_c^b \frac{|f(x)|^p}{|x-b|^{jp+s}}\,dx
$$

and hence for

$$I := \int_a^b \frac{|f(x)|^p}{\delta_{(a,b)}(x)^{jp+s}} \, dx.$$

It follows that for $j \in 2, \ldots k$,

$$\int_a^b \frac{|f^{(k-j)}(x)|^p}{\delta_{(a,b)}(x)^{jp+s}} \, dx \le \left(\frac{p}{jp+s-1} \right)^p \int_a^b \frac{|f^{(k-j+1)}(x)|^p}{\delta_{(a,b)}(x)^{(j-1)p+s}} \, dx, \quad (7.3.12)$$

and for $j = 1$,

$$\int_a^b \frac{|f^{(k-1)}(x)|^p}{\delta_{(a,b)}(x)^{p+s}} \, dx \le \left(\frac{p}{p+s-1} \right)^p \int_a^b |f^k(x)|^p \left\{ \frac{1}{(x-a)} + \frac{1}{(b-x)} \right\}^s \, dx. \tag{7.3.13}$$

Therefore

$$\int_a^b |f^k(x)|^p \left\{ \frac{1}{(x-a)} + \frac{1}{(b-x)} \right\}^s \, dx$$
$$\ge \left(\frac{p+s-1}{p} \right)^p \left(\frac{2p+s-1}{p} \right)^p \cdots \left(\frac{kp+s-1}{p} \right)^p \int_a^b \frac{|f(x)|^p}{\delta_{(a,b)}(x)^{kp+s}} \, dx$$

and the lemma is proved. $\qquad\qquad\qquad\qquad\qquad\qquad\qquad\qquad\square$

The following lemma is a key result in the proof of Theorem 7.7, and is a consequence of Lemma 7.8, and Theorems 2.1 and 2.6 in [130].

Lemma 7.9 *Let* $-\infty < a < b < \infty$ *and* $1 < \alpha < 2$. *Then for* $f \in C_0^\infty(a, b)$,

$$\int_{(a,b) \times (a,b)} \frac{|f'(x) - f'(y)|^2}{|x-y|^{1+\alpha}} \, dx \, dy \ge 2 \left(\frac{\alpha+1}{2} \right)^2 \kappa_{1,\alpha} \int_a^b \frac{|f(x)|^2}{\delta_{(a,b)}(x)^{\alpha+2}} \, dx. \tag{7.3.14}$$

For $1 < \alpha < p < \infty$,

$$\int_{(a,b) \times (a,b)} \frac{|f'(x) - f'(y)|^p}{|x-y|^{1+\alpha}} \, dx \, dy$$
$$\ge \left(\frac{\alpha+p-1}{p} \right)^p \mathcal{D}_{1,p,\alpha} \int_a^b \frac{|f(x)|^p}{|\delta_{(a,b)}(x)|^{\alpha+p}} \, dx. \tag{7.3.15}$$

Proof   From Theorem 2.1 in [130],

$$\int_{(a,b)\times(a,b)} \frac{|f'(x)-f'(y)|^2}{|x-y|^{1+\alpha}}\,dx\,dy$$

$$\geq 2\kappa_{1,\alpha}\int_a^b |f'(x)|^2\left[\frac{1}{x-a}+\frac{1}{b-x}\right]^\alpha dx$$

$$= 2\kappa_{1,\alpha}\left(\int_a^c+\int_c^b\right)|f'(x)|^2\left[\frac{1}{x-a}+\frac{1}{b-x}\right]^\alpha dx$$

$$\geq 2\kappa_{1,\alpha}\int_a^b \frac{|f'(x)|^2}{\delta_{(a,b)}(x)^\alpha}\,dx, \tag{7.3.16}$$

and (7.3.14) follows from the case $k=1$ of Lemma 7.8. The inequality (7.3.15) follows from Theorem 2.6 in [130] and Lemma 7.8.   □

The same argument as in the proof of (6.2.6) in Lemma 6.9, gives

Corollary 7.10   *Let J be an open subset of $\mathbb{R}$ and*

$$\delta_J(t) := \min\{|s| : t+s \notin J\}. \tag{7.3.17}$$

For $1 < \alpha < 2$ and $f \in C_0^\infty(J)$,

$$\int_{J\times J} \frac{|f'(x)-f'(y)|^2}{|x-y|^{1+\alpha}}\,dx\,dy \geq 2\left(\frac{\alpha+1}{2}\right)^2 \kappa_{1,\alpha}\int_J \frac{|f(x)|^2}{\delta_J(x)^{\alpha+2}}\,dx. \tag{7.3.18}$$

If $1 < \alpha < p < \infty$,

$$\int_{J\times J} \frac{|f'(x)-f'(y)|^p}{|x-y|^{1+\alpha}}\,dx\,dy \geq \left(\frac{\alpha+p-1}{p}\right)^p \mathcal{D}_{1,p,\alpha}\int_J \frac{|f(t)|^p}{\delta_J(t)^{p+\alpha}}\,dt. \tag{7.3.19}$$

Corollary 7.11   *For each x in the domain $\Omega \subset \mathbb{R}^n$ and $v \in \mathbb{S}^{n-1}$, define*

$$J(x,v) := \{t : x+tv \in \Omega\}, \tag{7.3.20}$$

$$\delta_{J(\mathbf{x},v)} := \min\{|t| : t \notin J(\mathbf{x},v)\}. \tag{7.3.21}$$

Let $1 < \alpha < p < \infty$ and set $D = (D_1, D_2, \ldots D_n)$, $D_i = \partial/\partial x_i$. Then for $x \in \Omega$, $f \in C_0^\infty(J(x,v))$ and $v \in \mathbb{S}^{n-1}$,

$$\int_{J(x,v)\times J(x,v)} \frac{|(Df\cdot v)(x+sv)-(Df\cdot v)(x+tv)|^p}{|s-t|^{1+\alpha}}\,ds\,dt$$

$$\geq E(\alpha,p)\int_{J(x,v)} |(Df\cdot v)(x+tv)|^p\frac{1}{\delta_{J(x,v)}(t)^\alpha}\,dt$$

$$\geq E(\alpha,p)\left(\frac{\alpha+p-1}{p}\right)^p \int_{J(x,v)} |f(x+tv)|^p\frac{1}{\delta_{J(x,v)}(t)^{\alpha+p}}\,dt, \tag{7.3.22}$$

where $E(\alpha,p) = \mathcal{D}_{1,p,\alpha}$ for $1 < p < \infty$ and $2\kappa_{1,\alpha}$ when $p=2$.

Proof Since Ω is an open connected set, then each $J(x, v)$ is an open set in $\mathbb{R}$. As a function of $t, f(x + tv) \in C_0^\infty(J(x, v))$ and by the chain rule,

$$\frac{d}{dt} f(x + tv) = (v \cdot Df)(x + tv).$$

Thus, (7.3.22) follows from Corollary 7.10 applied to $f(x + tv)$. $\qquad\square$

Lastly we need a lower bound for

$$e_p(n) := \left(\sum_{i=1}^n |v_i|^{p'}\right)^{-p/p'}, \quad v = (v_i) \in \mathbb{S}^{n-1}.$$

If $1 < p \le 2$,

$$\sum_{i=1}^n |v_i|^{p'} \le \sum_{i=1}^n |v_i|^2 = 1,$$

and if $p > 2$,

$$\sum_{i=1}^n |v_i|^{p'} \le \left(\sum_{i=1}^n |v_i|^2\right)^{p'/2} \left(\sum_{i=1}^n 1\right)^{1-p'/2} = n^{1-p'/2}.$$

Thus

$$\left(\sum_{i=1}^n |v_i|^{p'}\right)^{-p/p'} \ge \begin{cases} 1, & \text{if } 1 < p \le 2, \\ n^{-(p-2)/2}, & \text{if } p > 2. \end{cases}$$

$$(7.3.23)$$

Hence, for $v = (v_i) \in \mathbb{S}^{n-1}$,

$$|(v \cdot Df)(x + sv) - (v \cdot Df)(x + tv)|^p$$
$$= \left|\Sigma_{i=1}^n v_i D_i f(x + sv) - \Sigma_{i=1}^n v_i D_i f(x + tv)\right|^p$$
$$\le e_p(n)^{-1} \Sigma_{i=1}^n |D_i f(x + sv) - D_i f(x + tv)|^p,$$

and we have as in Lemma 6.10,

Lemma 7.12 *Let* $1/p < \sigma < 1$, $1 < p < \infty$ *and* $f \in C_0^\infty(\Omega)$. *Then*

$$\sum_{i=1}^n \int_\Omega \int_\Omega \frac{|D_i f(x) - D_i f(y)|^p}{|x - y|^{n+p\sigma}} \, dx \, dy$$

$$\ge \frac{\omega_{n-1}}{2} \int_{\mathbb{S}^{n-1}} e_p(n) \, d\omega(v) \int_{x:\, x\cdot v=0} d\mathcal{L}_v(x) \int_{x+sv\in\Omega} ds$$

$$\times \int_{x+tv\in\Omega} \left\{ \frac{|(v \cdot Df)(x + sv) - (v \cdot Df)(x + tv)|^p}{|s - t|^{1+p\sigma}} \right\} dt, \quad (7.3.24)$$

where $\mathcal{L}_v(x)$ denotes the $(n-1)$-dimensional Lebesgue measure on the plane $x \cdot v = 0$.

We now have all we need for the proof of Theorem 7.7.

Proof of Theorem 7.7   From Lemma 2.4 in [130],

$$\int_\Omega \int_\Omega \frac{|D_i f(x) - D_i f(y)|^2}{|x-y|^{n+2\sigma}} \, dx \, dy$$

$$= \frac{\omega_{n-1}}{2} \int_{S^{n-1}} d\omega(v) \int_{x:\, x \cdot v = 0} d\mathcal{L}_v(x) \int_{x+sv \in \Omega} ds$$

$$\times \int_{x+tv \in \Omega} \left\{ \frac{|D_i f(x+sv) - D_i f(x+tv)|^2}{|s-t|^{1+2\sigma}} \right\} dt.$$

Thus, on applying (7.3.24),

$$\sum_{i=1}^n \int_\Omega \int_\Omega \frac{|D_i f(x) - D_i f(y)|^2}{|x-y|^{n+2\sigma}} \, dx \, dy$$

$$\geq \frac{\omega_{n-1}}{2} \int_{S^{n-1}} d\omega(v) \int_{x:\, x \cdot v = 0} d\mathcal{L}_v(x) \int_{x+sv \in \Omega} ds$$

$$\times \int_{x+tv \in \Omega} \left\{ \frac{|(v \cdot Df)(x+sv) - (v \cdot Df)(x+tv)|^2}{|s-t|^{1+2\sigma}} \right\} dt.$$

From Corollary 7.11, we therefore have

$$\sum_{i=1}^n \int_\Omega \int_\Omega \frac{|D_i f(x) - D_i f(y)|^2}{|x-y|^{n+2\sigma}} \, dx \, dy$$

$$\geq \omega_{n-1} \, \kappa_{1,2\sigma} \int_{S^{n-1}} d\omega(v) \int_{x:\, x \cdot v = 0} d\mathcal{L}_v(x)$$

$$\times \int_{x+sv \in \Omega} |(v \cdot Df)(x+sv)|^2 \, \frac{1}{\delta_v(x+sv)^{2\sigma}} \, ds$$

$$\geq \left(\frac{2\sigma + 1}{2} \right)^2 \omega_{n-1} \, \kappa_{1,2\sigma} \int_{S^{n-1}} d\omega(v) \int_{x:\, x \cdot v = 0} d\mathcal{L}_v(x)$$

$$\times \int_{x+sv \in \Omega} |f(x+sv)|^2 \, \frac{1}{\delta_v^{2+2\sigma}(x+sv)} \, ds$$

$$\geq 2 \left(\frac{2\sigma + 1}{2} \right)^2 \kappa_{n,2\sigma} \int_\Omega \frac{|f(x)|^2}{M_{1+\sigma,2}(x)^{2+2\sigma}} \, dx.$$

The proof for $p = 2$ is complete. For general p the proof is similar.   □

7.4 Higher-Order Fractional Hardy–Rellich Inequalities

Some more notation and preliminary remarks are required before stating the main theorem.

For $v = (v_i) \in \mathbb{S}^{n-1}$, $\alpha = (\alpha_1, ..., \alpha_n) \in \mathbb{N}_0^n$ and $k \in \mathbb{N}$, use of the multinomial theorem shows that

$$(v \cdot D)^k \quad = \quad (v_1 D_1 + \cdots + v_n D_n)^k \qquad (7.4.1)$$

$$= \quad \sum_{|\alpha|=k} \frac{k!}{\alpha_1! ... \alpha_n!} (v_1 D_1)^{\alpha_1} ... (v_n D_n)^{\alpha_n}$$

$$:= \quad \sum_{|\alpha|=k} \frac{k!}{\alpha!} (v_\alpha D^\alpha), \quad v_\alpha = v_1^{\alpha_1} ... v_n^{\alpha_n}. \qquad (7.4.2)$$

For $p \in (1, \infty)$, $k \in \mathbb{N}$ and $v \in \mathbb{S}^{n-1}$, we shall need

$$S_{k,p'}(v) := \left(\sum_{|\alpha|=k} \left(\frac{k!}{\alpha!} \right)^{p'} |v_\alpha|^{p'} \right)^{1/p'}, \quad S_{k,p'} := \max_{v \in \mathbb{S}^{n-1}} S_{k,p'}(v), \qquad (7.4.3)$$

where $|v_\alpha|^2 := v_1^{2\alpha_1} + \cdots + v_n^{2\alpha_n}$.

Remark 7.13

1. It follows from (7.3.23) that

$$S_{1,p'}^p \leq \begin{cases} 1, & \text{if } 1 < p \leq 2, \\ n^{(p-2)/2(p-1)} & \text{if } 2 < p < \infty. \end{cases} \qquad (7.4.4)$$

2. Estimation of $S_{k,p'}$ when $k > 1$ requires more effort. For example, suppose that $k = 2$. Then there are two possibilities:

   (a) two components of α, say α_i and α_j, are 1 and the others are zero;
   (b) one component of α, say α_j, is 2 and the others are zero.

In case (a), $\alpha! = 1$ and $v_1^{2\alpha_1} + \cdots + v_n^{2\alpha_n} = v_i^2 + v_j^2 + n - 2$, so that $n - 2 \leq |v_\alpha|^2 \leq n - 1$ and

$$4(n-2) \leq \left(\frac{2}{\alpha!} \right)^2 |v_\alpha|^2 \leq 4(n-1).$$

In case (b), $\alpha! = 2$ and $n - 1 \leq |v_\alpha|^2 \leq v_j^2 + n - 1 \leq n$, showing that

$$n - 1 \leq \frac{2}{\alpha!} |v_\alpha|^2 \leq n.$$

In the sum for $S_{2,2}^2$ there are $n(n-1)/2$ terms of type (a) and n of type (b). Thus

$$2n(n-1)(2n-3) \leq S_{2,2}^2(v) \leq 2n(n-1)^2 + n^2,$$

and so

$$n(n-1)(2n-3) \leq S_{2,2}^2(v) \leq n(2n^2 - 3n + 2). \qquad (7.4.5)$$

The mean distance function for the higher-order inequalities is, for $1 < p < \infty$ and $1/p < \sigma < 1$,

$$\frac{1}{M_{k+\sigma,p}(x)^{p\sigma+kp}} = \frac{\sqrt{\pi}\,\Gamma\left(\frac{n+p\sigma+kp}{2}\right)}{\Gamma\left(\frac{1+p\sigma+kp}{2}\right)\Gamma\left(\frac{n}{2}\right)} \int_{S^{n-1}} \frac{1}{\delta_{v,\Omega}^{p\sigma+kp}(x)} d\omega(v), \qquad (7.4.6)$$

and the following constants are analogous to those in Section 7.3:

$$\mathcal{D}_{k,n,p,p\sigma} := \frac{2\pi^{(n-1)/2}\Gamma\left(\frac{1+pk+p\sigma}{2}\right)}{\Gamma\left(\frac{n+pk+p\sigma}{2}\right)}\mathcal{D}_{1,p,p\sigma}, \qquad (7.4.7)$$

$$\kappa_{k,n,2\sigma} = \frac{2\pi^{(n-1)/2}\Gamma\left(\frac{1+2k+2\sigma}{2}\right)}{\Gamma\left(\frac{n+2k+2\sigma}{2}\right)}\kappa_{1,2\sigma}, \qquad (7.4.8)$$

where $\mathcal{D}_{1,p,p\sigma}$ and $\kappa_{1,2\sigma}$ are given in (7.3.4) and (7.3.5). If Ω is convex with non-empty boundary, $0 < \sigma < 1$ and $1/\sigma < p < \infty$, then for all values of k, we have

$$M_{k+\sigma,p}(x) \leq \delta(x) := \inf\{|y - x| : y \notin \Omega\}. \qquad (7.4.9)$$

Note that in the case $k = 1$, our notation for (7.4.7) and (7.4.8) was $\mathcal{D}_{n,p,p\sigma}$ and $\kappa_{n,2\sigma}$. The constant

$$G(m\sigma, k, p) = \begin{cases} \Pi_{j=1}^{k}\left(\frac{jp+m\sigma-1}{p}\right)^p, & k \in \mathbb{N}, \\ 1, & k = 1 \end{cases} \qquad (7.4.10)$$

appears in our main theorem.

Theorem 7.14 *Let Ω be a domain in $\mathbb{R}^n$ with non-empty boundary, $1 < p < \infty$ and $1/p < \sigma < 1$. Then, for all $f \in C_0^{\infty}(\Omega)$,*

$$S_{k,p'}^p \sum_{|\alpha|=k} \int_{\Omega}\int_{\Omega} \frac{|(D^{\alpha}f(x) - D^{\alpha}f(y))|^p}{|x-y|^{n+p\sigma}}\,dx\,dy$$

$$= S_{k,p'}^p \sum_{j_1,\cdots,j_k=1}^{n} \int_{\Omega}\int_{\Omega} \frac{|(D_{j_1}\cdots D_{j_k}f)(x) - (D_{j_1}\cdots D_{j_k}f)(y)|^p}{|x-y|^{n+p\sigma}}\,dx\,dy$$

$$\geq E_{k,n,p,p\sigma}G(p\sigma, k, p) \int_{\Omega} \frac{|f(x)|^p}{M_{k+\sigma,p}(x)^{p\sigma+kp}}\,dx, \qquad (7.4.11)$$

where $E_{k,n,p,p\sigma} = \mathcal{D}_{k,n,p,p\sigma}$; when $p = 2$ the inequality holds with $E_{k,n,2,2\sigma} = 2\kappa_{k,n,2\sigma}$.

Proof Corollaries 7.10 and 7.11 have the following analogues for any $k \in \mathbb{N}_0$:

$$\int_{J \times J} \frac{|f^{(k)}(x) - f^{(k)}(y)|^p}{|x-y|^{1+p\sigma}} \, dx \, dy$$
$$\geq E(p\sigma, p) G(p\sigma, k, p) \int_J \frac{|f(x)|^p}{|\delta_J(x)|^{kp+p\sigma}} \, dx,$$

where $E(p\sigma, p) = \mathcal{D}_{1,p,p\sigma}$, $E(2\sigma, 2) = \kappa_{1,2\sigma}$, and

$$\int_{J(x,v) \times J(x,v)} \frac{|(v \cdot D)^k f(x+sv) - (v \cdot D)^k f(x+tv)|^p}{|s-t|^{1+p\sigma}} \, ds \, dt$$
$$\geq E(p\sigma, p) G(p\sigma, k, p) \int_{J(x,v)} \frac{|f(x+tv)|^p}{\delta_{J(x,v)}(t)^{kp+p\sigma}} \, dt.$$

To proceed with the proof, we need the following inequality to obtain an analogue of Lemma 7.12, and thus of Lemma 2.4 in [130]. From (7.4.1) and (7.4.2),

$$(v \cdot D)^k = \sum_{|\alpha|=k} \frac{k!}{\alpha!} v_\alpha D^\alpha$$

for $v_\alpha = v_1^{\alpha_1} \cdots v_n^{\alpha_n}$, and

$$\left| ((v \cdot D)^k f)(x+sv) - ((v \cdot D)^k f)(x+tv) \right|^p$$
$$= \left| \sum_{|\alpha|=k} \frac{k!}{\alpha!} \{(v_\alpha \cdot D^\alpha)f(x+sv) - (v_\alpha \cdot D^\alpha)f(x+tv)\} \right|^p$$
$$\leq \left(\sum_{|\alpha|=k} \left(\frac{k!}{\alpha!}\right)^{p'} |v_\alpha|^{p'} \right)^{p/p'} \left(\sum_{|\alpha|=k} |(D^\alpha f)(x+sv) - (D^\alpha f)(x+tv)|^p \right)$$
$$\leq S_{k,p'}^p \left(\sum_{|\alpha|=k} |(D^\alpha f)(x+sv) - (D^\alpha f)(x+tv)|^p \right),$$

where $S_{k,p'}$ is defined in (7.4.3). Then, for $1/p < \sigma < 1$, $1 < p < \infty$ and $f \in C_0^\infty(\Omega)$,

$$S_{k,p'}^p \int_\Omega \int_\Omega \frac{\sum_{|\alpha|=k} |(D^\alpha f)(x) - (D^\alpha f)(y)|^p}{|x-y|^{n+p\sigma}} \, dx \, dy$$
$$\geq \frac{\omega_{n-1}}{2} \int_{S^{n-1}} d\omega(v) \int_{x: x \cdot v = 0} d\mathcal{L}_v(x) \int_{x+sv \in \Omega} ds$$
$$\times \int_{x+tv \in \Omega} \left\{ \frac{|(v \cdot D)^k f(x+sv) - (v \cdot D)^k f(x+tv)|^p}{|s-t|^{1+p\sigma}} \right\} dt,$$

where $\mathcal{L}_\nu(x)$ denotes the $(n-1)$-dimensional Lebesgue measure on the plane $x \cdot \nu = 0$. It follows that

$$
\begin{aligned}
S_{k,p'}^p \int_\Omega \int_\Omega & \frac{\sum_{|\alpha|=k} |(D^\alpha f)(x) - (D^\alpha f)(y)|^p}{|x-y|^{n+p\sigma}} \, dx \, dy \\
&\geq E(p\sigma, p) G(p\sigma, k, p) \int_{S^{n-1}} d\omega(\nu) \int_{x: \, x\cdot\nu=0} d\mathcal{L}_\nu(x) \\
&\quad \times \int_{x+s\nu\in\Omega} \frac{|f(x+s\nu)|^p}{\delta_\nu(x+s\nu)^{p\sigma+kp}} \, ds \\
&\geq E(p\sigma, p) G(p\sigma, k, p) \int_{S^{n-1}} d\omega(\nu) \int_\Omega \frac{|f(x)|^p}{\delta_{\nu,\Omega}(x)^{p\sigma+kp}} \, dx \\
&\geq E_{k,n,p,p\sigma} \, G(p\sigma, k, p) \int_\Omega \frac{|f(x)|^p}{M_{k+\sigma,p}(x)^{p\sigma+kp}} \, dx.
\end{aligned}
$$

This completes the proof. $\qquad\qquad\qquad\qquad\qquad\qquad\qquad\qquad\qquad\qquad$ □

7.5 Higher-Order Inequality with a Remainder

An analogue of Proposition 6.14 is now readily established for higher-order Hardy–Rellich inequalities. First we note that Corollary 7.11 has the extension

Corollary 7.15 *Let Ω be a bounded domain in $\mathbb{R}^n$, and for $x \in \Omega$ and $\nu \in \mathbb{S}^{n-1}$, define*

$$
J(x, \nu) := \{t: x + t\nu \in \Omega\},
$$
$$
\delta_{J(x,\nu)} := \min\{|t|: t \notin J(x, \nu)\}.
$$

Then for $1/2 < \sigma < 1, f \in C_0^\infty(\Omega)$ and $k \in \mathbb{N}_0$,

$$
\begin{aligned}
\int_{J(x,\nu)\times J(x,\nu)} & \frac{|(\nu \cdot D)^k f(x+r\nu) - (\nu \cdot D)^k f(x+t\nu)|^2}{|r-t|^{1+2\sigma}} \, dr \, dt \\
&\geq 2\kappa_{1,2\sigma} G(2\sigma, k, 2) \int_{J(x,\nu)} \frac{|f(x+t\nu)|^2}{\delta_{J(x,\nu)}(t)^{2k+2\sigma}} \, dt \\
&\quad + 2\frac{4 - 2^{3-2\sigma}}{2\sigma \, \text{diam}\,(J(x,\nu))} G(2\sigma - 1, k, 2) \int_{J(x,\nu)} \frac{|f(x)|^2}{|\delta_J(x)|^{2k+2\sigma-1}} \, dx, \quad (7.5.1)
\end{aligned}
$$

using (7.4.8).

We may now follow a similar argument to that in the proof of Theorem 7.14, and in Dyda's Theorem 6.13, to derive

Theorem 7.16 *Let Ω be a bounded domain in $\mathbb{R}^n$ with non-empty boundary and $k \in \mathbb{N}_0$, $1/2 < \sigma < 1$. Then, for all $f \in C_0^\infty(\Omega)$,*

$$
S_{k,2}^2 \sum_{|\alpha|=k} \int_\Omega \int_\Omega \frac{|(D^\alpha f(x) - D^\alpha f(y))|^2}{|x-y|^{n+2\sigma}} \, dx \, dy
$$

$$
\geq 2\kappa_{k,n,2\sigma} G(2\sigma, k, 2) \int_\Omega \frac{|f(x)|^p}{M_{k+\sigma,2}(x)^{2\sigma+2k}} \, dx
$$

$$
+ 2 \frac{4 - 2^{3-2\sigma}}{2\sigma \operatorname{diam}(\Omega)} \frac{\kappa_{k,n,2\sigma-1}}{\kappa_{1,2\sigma-1}} G(2\sigma - 1, k, 2) \int_\Omega \frac{|f(x)|^2}{M_{k+\sigma-1/2,2}(x)^{2k+2\sigma-1}} \, dx,
$$

$$
(7.5.2)
$$

where, by (7.4.8),

$$
\frac{\kappa_{k,n,2\sigma-1}}{\kappa_{1,2\sigma-1}} = \frac{2\pi^{(n-1)/2} \Gamma\left(\frac{2k+2\sigma}{2}\right)}{\Gamma\left(\frac{n+2k+2\sigma-1}{2}\right)}
$$

and $M_{k+\alpha,2}$ is defined in (6.2.1). If Ω is convex, $M_{k+\alpha,2}(x) \leq \delta(x) := \inf\{|y - x| : y \notin \Omega\}$.

The constant multiplying the first integral on the right-hand side of (7.5.2) cannot be replaced by a larger one in the case $k = 0$, but this is not proved for $k \geq 1$.

7.6 Higher-Order Classical Inequalities

It is proved by Bourgain, Brezis and Mironescu in [23] that if Ω is a connected open subset of $\mathbb{R}^n$ and $1 < p < \infty$, then for all $f \in C_0^\infty(\Omega)$,

$$
\lim_{\sigma \to 1^-} (1 - \sigma) \int_\Omega \int_\Omega \frac{|f(x) - f(y)|^p}{|x-y|^{n+p\sigma}} \, dx \, dy = K(n, p) \int_\Omega |\nabla f(x)|^p \, dx
$$

for some positive constant $K(n, p)$ depending only on n and p; see Corollary 3.20 and Remark 3.21. If $p = 2$, the following precise information is established in [78], Lemma 3.1:

$$
\int_{\mathbb{R}^n} \int_{\mathbb{R}^n} \frac{|f(x) - f(y)|^2}{|x-y|^{n+2\sigma}} \, dx \, dy = 2C(n, \sigma)^{-1} \int_{\mathbb{R}^n} \left|(-\Delta)^{\sigma/2} f(x)\right|^2 dx \quad (7.6.1)
$$

for $0 < \sigma < 1$ and

$$
\frac{1}{2} C(n, \sigma) = 2^{2\sigma-1} \pi^{-n/2} \frac{\Gamma\left(\frac{n}{2} + \sigma\right)}{|\Gamma(-\sigma)|}. \quad (7.6.2)
$$

In (6.6.1), $(-\Delta)^{\sigma/2} f(x) := \left[F^{-1} \left(|\xi|^\sigma \hat{f}(\xi) \right) \right](x)$, where $\hat{f} = F(f)$, and it follows that

$$\sum_{|\alpha|=k} \int_{\mathbb{R}^n} \int_{\mathbb{R}^n} \frac{|(D^\alpha f(x) - D^\alpha f(y))|^2}{|x-y|^{n+2\sigma}} \, dx\, dy$$

$$= 2C(n,\sigma)^{-1} \sum_{|\alpha|=k} \int_{\mathbb{R}^n} \left| (-\Delta)^{\sigma/2} D^\alpha f(x) \right|^2 dx$$

$$= 2C(n,\sigma)^{-1} \sum_{|\alpha|=k} \int_{\mathbb{R}^n} \left| (|\xi|^2)^{\sigma/2} (i\xi)^\alpha \hat{f}(\xi) \right|^2 d\xi$$

$$= 2C(n,\sigma)^{-1} \int_{\mathbb{R}^n} \left| (-\Delta)^{\frac{\sigma+k}{2}} f(x) \right|^2 dx.$$

Hence, for $f \in C_0^\infty(\Omega)$,

$$\int_{\mathbb{R}^n} \left| (-\Delta)^{\frac{\sigma+k}{2}} f(x) \right|^2 dx$$

$$= \frac{1}{2} C(n,\sigma) \sum_{|\alpha|=k} \int_{\mathbb{R}^n} \int_{\mathbb{R}^n} \frac{|(D^\alpha f(x) - D^\alpha f(y))|^2}{|x-y|^{n+2\sigma}} \, dx\, dy$$

$$\geq \frac{1}{2} C(n,\sigma) \sum_{|\alpha|=k} \int_\Omega \int_\Omega \frac{|(D^\alpha f(x) - D^\alpha f(y))|^2}{|x-y|^{n+2\sigma}} \, dx\, dy. \qquad (7.6.3)$$

In (7.4.11), the constant multiple of the integral on the right-hand side is

$$2G(2\sigma, k, 2)\kappa_{n,2\sigma}$$

in which, as $\sigma \to 1-$, $G(2, k, 2) = \left(\frac{(2k+1)!}{k! 2^{2k+1}} \right)^2 = \left(\frac{2}{\sqrt{\pi}} \Gamma \left(k + \frac{3}{2} \right) \right)^2$ and $\kappa_{n,2\sigma}$ is asymptotic to

$$\frac{\pi^{(n-1)/2} \Gamma \left(\frac{3+2k}{2} \right)}{\Gamma \left(\frac{n+2k+2}{2} \right)} \frac{2^{-2}}{\sqrt{\pi}} \Gamma(3/2)(1-\sigma)^{-1}$$

$$= \frac{1}{8} \pi^{(n-1)/2} \frac{\Gamma \left(\frac{3+2k}{2} \right)}{\Gamma \left(\frac{n+2k+2}{2} \right)} (1-\sigma)^{-1}. \qquad (7.6.4)$$

Also, as $\sigma \to 1-$, $\frac{1}{2} C(n,\sigma)$ in (7.6.3) satisfies

$$\frac{1}{2} C(n,\sigma) \sim 2\pi^{-n/2} \Gamma(n/2+1)(1-\sigma), \qquad (7.6.5)$$

and for $f \in C_0^\infty(\Omega)$,

$$I := \lim_{\sigma \to 1-} \int_{\mathbb{R}^n} \left| (-\Delta)^{(\sigma+k)/2} f(x) \right|^2 dx = \int_{\mathbb{R}^n} \left| (-\Delta)^{(1+k)/2} f(x) \right|^2 dx.$$

This follows by dominated convergence, on noting that

$$I = \lim_{\sigma \to 1-} \int_{\mathbb{R}^n} \left| (|\xi|^2)^{(\sigma+k)/2} \hat{f}(\xi) \right|^2 d\xi,$$

and, for $0 \le \sigma \le 1$,

$$\left| (|\xi|^2)^{(\sigma+k)/2} \hat{f}(\xi) \right|^2 \le \left| \left[(|\xi|^2)^{(1+k)/2} + 1 \right] \hat{f}(\xi) \right|^2$$
$$= \left| F\left(\left[(-\Delta)^{(1+k)/2} + 1 \right] f \right)(\xi) \right|^2 \in L_1(\mathbb{R}^n).$$

Hence from (7.4.11) and (7.6.3), the inequality we get in the limit as $\sigma \to 1-$ is

$$\int_{\mathbb{R}^n} \left| \Delta^{\frac{1+k}{2}} f(x) \right|^2 dx \ge K(n,k) \lim_{\sigma \to 1-} \int_\Omega \frac{|f(x)|^2}{M_{k+\sigma,2}(x)^{2k+2\sigma}} dx, \tag{7.6.6}$$

where

$$K(n,k) = 2 S_{k,2}^{-2} \frac{\left[\Gamma\left(\frac{3+2k}{2} \right) \right]^3 \Gamma\left(\frac{n}{2} + 1 \right)}{\pi^{3/2} \Gamma\left(\frac{n+2k+2}{2} \right)}. \tag{7.6.7}$$

As $\sigma \to 1-$, $M_{k+\sigma,2}^{-2k-2\sigma}$ is bounded by $M_{k+1,2}^{-2k-2}$ on the support of f. Since $|f|^2 M_{k+1,2}^{-2k-2} \in L_1(\Omega)$ is proved in [143] (see (7.6.10) below) it follows by dominated convergence from (7.6.6) that

$$\int_{\mathbb{R}^n} \left| \Delta^{\frac{1+k}{2}} f(x) \right|^2 dx \ge K(n,k) \int_\Omega \frac{|f(x)|^2}{M_{k+1,2}(x)^{2k+2}} dx. \tag{7.6.8}$$

In [143], Owen establishes the following Hardy–Rellich inequality for polyharmonic operators with a sharp constant:

$$\int_\Omega \bar{f}(x) \left[(-\Delta)^m f \right](x) dx \ge \frac{\Gamma\left(\frac{n}{2} + m \right) \Gamma\left(m + \frac{1}{2} \right)}{\Gamma\left(\frac{n}{2} \right) \Gamma\left(\frac{1}{2} \right)} \int_\Omega \frac{|f(x)|^2}{a_m^{2m}(x)} dx$$

for all $f \in C_0^\infty(\Omega)$, $m \in \mathbb{N}$, and where

$$\frac{1}{a_m^{2m}(x)} = \int_{\mathbb{S}^{n-1}} \frac{1}{\delta_\nu(x)^{2m}} d\omega(\nu).$$

Owen expresses his result in the quadratic form sense

$$((-\Delta)^m f, f) \ge \frac{\Gamma\left(\frac{n}{2} + m \right) \Gamma\left(m + \frac{1}{2} \right)}{\Gamma\left(\frac{n}{2} \right) \Gamma\left(\frac{1}{2} \right)} (Af, f),$$

where $(\cdot, \cdot)$ is the $L_2(\Omega)$ inner-product, $(-\Delta)^m$ is the polyharmonic operator of order $2m$ and A is the operator of multiplication by $1/a_m^{2m}(x)$. On $C_0^\infty(\Omega)$,

$(-\Delta)^m$ is the restriction of $F^{-1}(|\cdot|^{2m})$, where F is the Fourier transform. In our notation, with $m = k + 1$,

$$\frac{1}{a_m^{2m}(x)} = \frac{\Gamma\left(k + \frac{3}{2}\right)\Gamma\left(\frac{n}{2}\right)}{\sqrt{\pi}\,\Gamma\left(\frac{n}{2} + k + 1\right)} \frac{1}{M_{k,1,2,\Omega}(x)^{2+2k}}.$$

Owen's inequality therefore implies

$$\int_\Omega \bar{f}(x)\left[(-\Delta)^{k+1}f\right](x)\,dx \geq K_0(k) \int_\Omega \frac{|f(x)|^2}{M_{k+1,2}(x)^{2+2k}}\,dx, \tag{7.6.9}$$

where

$$K_0(k) = \frac{\left[\Gamma\left(\frac{3+2k}{2}\right)\right]^2}{\pi}. \tag{7.6.10}$$

Hence, in particular, $|f|^2 M_{k+1,2}^{-2-2k} \in L_1(\mathbb{R}^n)$, as noted earlier.

The inequality (7.6.6) can also be expressed in the form sense, namely

$$\int_\Omega \bar{f}(x)\left[(-\Delta)^{k+1}f\right](x)\,dx \geq K(n,k) \int_\Omega \frac{|f(x)|^2}{M_{k+1,2}(x)^{2+2k}}\,dx,$$

and we have from (7.6.7) and (7.6.10),

$$\frac{K_0(k)}{K(n,k)} = S_{k,2}^2 \frac{\sqrt{\pi}\,\Gamma\left(\frac{n+2k+2}{2}\right)}{2\Gamma\left(\frac{3+2k}{2}\right)\Gamma\left(\frac{n}{2} + 1\right)}. \tag{7.6.11}$$

When $k = 0$ we have $K(n,0) = K_0(0) = 1/4$, which confirms that the constant in the Hardy case of Corollary 7.11 with $p = 2$ is sharp, as already proved in [130]. However we cannot claim this for $k \geq 1$; for instance, when $k = 1$, the value $K_0(1) = 9/16$ is sharp, but $K(n,1) = (9/16)(3/(n+2)) < K_0(1)$ for $n > 1$.

When $p \neq 2$ it seems harder to use Fourier transform techniques. However, there is an analogue of Corollary 1.4.8 of [63] that can be established by induction, namely that if $p \in (1, \infty)$ and $m \in \mathbb{N}$, then for all $\alpha \in \mathbb{N}_0^n$ with $|\alpha| = 2m$,

$$\left\|D^\alpha f|L_p\left(\mathbb{R}^n\right)\right\| \leq c_p^m \left\|\Delta^m f|L_p\left(\mathbb{R}^n\right)\right\| \tag{7.6.12}$$

for all smooth f with compact support, where

$$c_p = \cot^2\left(\frac{\pi}{2p^*}\right), p^* = \max(p, p'). \tag{7.6.13}$$

This enables higher-order counterparts of Theorem 7.4 to be proved.

References

[1] Abramovich, Y. A. and Aliprantis, C. D., An invitation to operator theory, *Graduate Studies in Mathematics*, Amer. Math. Soc. **50**, Providence, RI, 2002.

[2] Adams, R. A., *Sobolev Spaces*, Academic Press, New York, 1975.

[3] Adimurti, A. and Tintarev, K., Hardy inequalities for weighted Dirac operators, Ann. Mat. Pura Appl. **189** (2010), 241–251.

[4] Avkhadiev, F. G., Hardy type inequalities in higher dimensions with explicit estimate of constants, Lobachesvskii J. Math. **21** (2006), 3–31.

[5] Avkhadiev, F. G. and Laptev, A., Hardy inequalities for non-convex domains. In *Around the Research of Vladimir Maz'ya, I* (F. Laptev, ed.), International Mathematical Series **11**, pp. 1–12, Springer, New York (2010).

[6] Avkhadiev, F. G. and Wirths, K.-J., Sharp Hardy type inequalities with Lamb's constants, Bull. Belg. Math. Soc., Simon Stevin **18** (2011), 723–736.

[7] Avkhadiev, F. G., Families of domains with best possible Hardy constants, Russian (Iz VUZ) **57** (2013), 49–52.

[8] Avkhadiev, F. G. and Makarov, R. V., Hardy type inequalities on domains with complex complement and uncertainty principle of Heisenberg, Lobachevskii J. Math. **40**(9) (2019), 1250–1259.

[9] Almgren, F. J. and Lieb, E.H., Symmetric decreasing rearrangement is sometimes continuous, J. Amer. Math. Soc. **2**(2) (1989), 683–773.

[10] Alvino, A., Sulla diseguaglianza di Sobolev in spazi di Lorentz, Boll. Un. Mat. Ital. A **14** (1977), 148–156.

[11] Alvino, A., Lions, P.-L. and Trombetti, G., On optimization problems with prescribed rearrangements, Nonlinear Anal. **13** (1989), 185–220.

[12] Ancona, A., On strong barriers and an inequality of Hardy for domains in $\mathbb{R}^2$, J. Lond. Math. Soc. **33** (1986), 274–290.

[13] Aronszajn, N., Boundary values of functions with finite Dirichlet integral, Tech. Report 14, Univ. of Kansas, 1955, 77–94.

[14] Balinsky, A., Evans, W. D., Hundertmark, D. and Lewis, R. T., On inequalities of Hardy-Sobolev type, Banach J. Math. Anal. **2**(2) (2008), 94–106.

[15] Balinsky, A., Evans, W. D. and Lewis, R. T., *The Analysis and Geometry of Hardy's Inequality*, Springer, 2015.

[16] Bañuelos, R. and Wang, G., Sharp inequalities for martingales with applications to the Beurling-Ahlfors and Riesz transformations, Duke Math. J. **80** (1995), 575–600.

[17] Barbatis, G., Filippas, S. and Tertikas, A., A unified approach to improved L^p Hardy inequalities with best constants, Trans. Amer. Math. Soc. **356**(6) (2004), 2169–2196.

[18] Barbatis, G. and Tertikas, A., On the Hardy constant of non-convex planar domains: the case of a quadrilateral, J. Funct. Anal. **266** (2014), 3701–3725.

[19] Barbatis, G. and Tertikas, A., On the Hardy constant of some non-convex planar domains, in *Geometric Methods in PDEs*. Springer INdAM, vol. 13 (2015), 15–41.

[20] Benguria, R. D., Frank, R. L. and Loss, M., The sharp constant in the Hardy–Sobolev–Maz'ya inequality in three dimensional upper half-space, Math. Res. Lett. **15**(4) (2008), 613–622.

[21] Bennett, C. and Sharpley, R., *Interpolation of Operators*, Academic Press, 1988.

[22] Bogdan, D. and Dyda, B., The best constant in a fractional Hardy inequality, Math. Nachr. **284**(5–6) (2011), 629–638.

[23] Bourgain, J., Brezis, H. and Mironescu, P., Another look at Sobolev spaces. In *Optimal Control and Partial Differential Equations (In Honor of Professor Alain Bensoussan's 60th Birthday)* (J. L. Menaldi et al., eds.), pp. 439–455, IOS Press, Amsterdam (2001).

[24] Bourgain, J., Brezis, H. and Mironescu, P., Limiting embedding theorems for $W^{s,p}$ when $s \uparrow 1$ and applications, J. d'Analyse Math. **87** (2002), 77–101.

[25] Brasco, L. and Cinti, E., On fractional Hardy inequalities in convex sets, Discr. Contin. Dyn. Syst. Ser. A **18** (2018), 4019–4040.

[26] Brasco, L. and Franzina, G., On the Hong-Krahn-Szegö inequality for the p-Laplace operator, Manuscripta Math. **142** (2013), 537–557.

[27] Brasco, L. and Franzina, G., Convexity properties of Dirichlet integrals and Picone-type inequalities, Kodai Math. J. **37** (2014), 769–799.

[28] Brasco, L., Lindgren, E. and Parini, E., The fractional Cheeger problem, Interfaces and Free Boundaries **16** (2014), 419–458.

[29] Brasco, L. and Parini, E., The second eigenvalue of the fractional p-Laplacian, Adv. Calc. Var. **9** (2016), 323–355.

[30] Brasco, L., Parini, E. and Squassina, M., Stability of variational eigenvalues for the fractonal p-Laplacian, Discr. Cont. Dyn. Syst. **36** (2016), 1813–1845.

[31] Brasco, L. and Salort, A., A note on homogeneous Sobolev spaces of fractional order, Annali di Math. Pura ed Appl. **198** (2019), 1295–1330.

[32] Brezis, H., *Functional Analysis, Sobolev Spaces and Partial Differential Equations,* Springer, New York, 2011.

[33] Brezis, H., How to recognise constant functions, Russ. Math. Surv. **57** (2002), 693–708.

[34] Brezis, H. and Marcus, M., Hardy's inequalities revisited, Ann. Scuola Norm. Sup. Pisa Cl. Sci. **25**(4) (1997), 217–237.

[35] Brezis, H., Schaftingen, J. V. and Yung, Po-Lam, Going to Lorentz when fractional Sobolev, Gagliardo and Nirenberg estimates fail, Calc. Var. **60**(129) (2021), 1–12.

[36] Brezis, H., Schaftingen, J. V. and Yung, Po-Lam, A surprising formula for Sobolev norms, Proc. Natl. Acad. Sci. USA **118**(8), e2025254118(2021), https://doi.org/10.1073/pnas.202524118.

[37] Browder, F. E., Nonlinear elliptic boundary value problems and the generalised topological degree, Bull. Amer. Math. Soc. **76** (1970), 999–1005.

[38] Buncur, C. and Valdinoci, E., *Nonlocal Diffusion and Applications*, Lecture Notes of the Unione Matematica Italiana, **20**, Springer, Cham: Unione Matematica Italiana, Bologna, 2016.

[39] Bunt, L. H. N., *Bijdrage tot de Theorie der convexe Puntverzamelingen*. Thesis, University of Groningen, Amsterdam (1934).

[40] Carbotti, A., Dipierro, S., and Valdinoci, E. *Local Density of Solutions to Fractional Equations*, De Gruyter Studies in Mathematics, **74**, De Gruyter, Berlin, 2019.

[41] Cassani, D., Ruf, B. and Tarsi, C., Equivalent and attained version of Hardy's inequality in $\mathbb{R}^n$, J. Funct. Anal. **275** (2018), 3301–3324.

[42] Chowdhury, I., Csaló, G., Roy, P. and Sk, F., Study of fractional Poincaré inequalities on unbounded domains, Discr. Contin. Dyn. Syst. **41**(6) (2021), 2993–3020.

[43] Chen, B.-Y., Hardy-type inequalities and principal frequency of the p-Laplacian, arXiv: 2007.06782v1.

[44] Chen, Z.-Q. and Song, R., Hardy inequality for censored stable processes, Tohoku Math. J. **55** (2003), 439–450.

[45] Cianchi, A., Quantitative Sobolev and Hardy inequalities, and related symmetrization principles. In *Sobolev Spaces in Mathematics I* (V. Maz'ya, ed.), International Mathematical Series, pp. 87–116, Springer, New York (2009).

[46] Cuesta, M., Minimax theorems on C^1 manifolds via Ekeland variational principle, Abstr. Appl. Anal. **13** (2003), 757–768.

[47] Davies, E. B., Some norm bounds and quadratic form inequalities for Schrödinger operators (II), J. Operator Theory **12** (1984), 177–196.

[48] Davies, E. B., The Hardy constant, Quart. J. Math. Oxford (2) **46** (1994), 417–431.

[49] Davies, E. B. and Hinz, A., Explicit constants for Rellich inequalities in $L^p(\Omega)$, Math. Zeit. **227**(3) (1998), 511–523.

[50] Di Castro, A., Kuusi, T. and Palatucci, G., Local behavior of fractional $p-$minimisers, Ann. Inst. H. Poincaré Anal. Non Linéaire **33**(5) (2015), 1279–1299.

[51] Drábek, P. and Milota, J., *Methods of Nonlinear Analysis: Applications to Differential Equations*, Birkhäuser, Basel, 2007.

[52] Drelichman, I. and Durán, R. G., Improved Poincaré inequalities in fractional Sobolev spaces, Ann. Acad. Sci. Fennicae Math. **43** (2018), 885–903.

[53] Dubinskii, Yu. A., A Hardy-type inequality and its applications, Proc. Steklov Inst. Math. **269** (2010), 106–126.

[54] Dunford, N. and Schwartz, J. T., *Linear Operators I*, Interscience, New York and London, 1958.

[55] Dyda, B., A fractional order Hardy inequality, Illinois J. Math. **48** (2004), 575–588.

[56] Dyda, B., Fractional Hardy inequality with remainder term, Coll. Math. **122** (1) (2011), 59–67.

[57] Dyda, B. and Frank, R. L., Fractional Hardy–Sobolev–Maz'ya inequality for domains, Studia Math. **208** (2012), 151–166.

[58] Dyda, B., Ihnatsyeva, L. and Vähäkangas, A. V., On improved fractional Sobolev-Poincaré inequality, Ark. Mat. **54** (2016), 437–454.

[59] Dyda, B., Lehrbäck, J. and Vähäkangas, A. V., Fractional Hardy-Sobolev type inequalities for half spaces and John domains, Proc. Amer. Math. Soc. **140**(8) (2018), 3393–3402.

[60] Edmunds, D. E. and Evans, W. D., *Hardy Operators, Function Spaces and Embeddings*, Springer Monographs in Mathematics, Springer, Berlin/Heidelberg/New York, 2004.

[61] Edmunds, D. E. and Evans, W. D., *Representations of Linear Operators between Banach Spaces*, Birkhäuser, Basel, 2013.

[62] Edmunds, D. E., Hurri-Syrjänen, R. and Vähäkangas, A. V., Fractional Hardy-type inequalities in domains with uniformly fat complement, Proc. Amer. Math. Soc. **142**(3) (2014), 897–907.

[63] Edmunds, D. E. and Evans, W. D., The Rellich inequality, Rev. Mat. Complut. **29**(3) (2016), 511–530.

[64] Edmunds, D. E. and Evans, W. D., *Spectral Theory and Differential Operators*, 2nd edition, Oxford University Press, Oxford, 2018.

[65] Edmunds, D. E. and Evans, W. D., *Elliptic Differential Operators and Spectral Analysis*, Springer Nature, Switzerland AG, 2018.

[66] Edmunds, D. E., Evans, W. D. and Lewis, R. T., Fractional inequalities of Rellich type, Pure Appl. Funct. Anal., to appear.

[67] Edmunds, D. E., Gogatishvili, A. and Nekvinda, A., Almost compact and compact embeddings of variable exponent spaces, Studia Math., to appear.

[68] Edmunds, D. E., Lang, J., Non-compact embeddings of Sobolev spaces, to appear.

[69] Edmunds, D. E., Lang, J. and Mihula, Z., Measure of noncompactness of Sobolev embeddings on strip-like domains, J. Approx. Theory **269** (2021), 105608.

[70] Edmunds, D. E. and Triebel, H., *Function Spaces, Entropy Numbers, Differential Operators*, Cambridge University Press, Cambridge, 1996.

[71] Ekeland, I., On the variational principle, J. Math. Anal. Appl. **47** (1974), 324–353.

[72] Evans, W. D. and Harris, D. J., Sobolev embeddings for generalized ridged domains, Proc. Lond. Math. Soc. **54** (1987), 141–175.

[73] Evans, W. D. and Lewis, R. T., On the Rellich inequality with magnetic potentials, Math. Zeit. **251** (2005), 267–284.

[74] Evans, W. D. and Lewis, R. T., Hardy and Rellich inequalities with remainders, J. Math. Inequalities **1**(4) (2007), 473–490.

[75] Evans, W. D. and Schmidt, K. M., A discrete Hardy–Laptev–Weidl inequality and associated Schrödinger-type operators, Rev. Mat. Complut. **22**(1) (2009), 75–90.

[76] Fernández-Martínez, P., Manzano, A. and Pustylnik, E., Absolutely continuous embeddings of rearrangement-invariant spaces, Mediterr. J. Math. **7** (2010), 539–552.

[77] Filippas, S., Maz'ya, V., and Tertikas, A., Critical Hardy-Sobolev inequalities, J. Math. Pures. Appl. **87** (2007), 37–56.

[78] Frank, R. L., Lieb, E. and Seiringer, R., Hardy–Littlewood–Thirring inequalities for fractional Schrödinger operators, J. Amer. Math. Soc. **21**(4) (2008), 925–950.

[79] Frank, R. L., Eigenvalue bounds for Laplacians and Schrödinger operators: a review, arXiv:1603.09736v2 4 Nov. (2017), 1–24.

[80] Frank, R. L., and Larsen, S., Two consequences of Davies' Hardy inequality, Funksional Anal. i Prilozhen **55** (2021), 118–121.

[81] Frank, R. L. and Loss, M., Hardy–Sobolev–Maz'ya inequalities for arbitrary domains, J. Math. Pures Appl. **97** (2012), 39–54.

[82] Frank, R. L. and Seiringer, R., Non-linear ground state representations and sharp Hardy inequalities, J. Functional Anal. **255** (2008), 3407–3430.

[83] Frank, R. L. and Seiringer, R., Sharp fractional Hardy inequalities in half-spaces. In *Around the Research of Vladimir Maz'ya, I* (F. Laptev, ed.), International Mathematical Series **11**, pp. 161–167, Springer, New York (2010).

[84] Franzina, G. and Palatucci, G., Fractional p−eigenvalues, Riv. Mat. Univ. Parma **5**(2) (2014), 315–328.

[85] Fremlin, D. H., Skeletons and central sets, Proc. Lond. Math. Soc. **74** (1997), 701–720.

[86] Gagliardo, E., Proprietà di alcune classi di funzioni in più variabili, Ricerche Mat. **7** (1958), 102–137.

[87] Gesztesy, F., Pang, M. H. and Stanfill, J., Bessel-type operators and a refinement of Hardy's inequality, arXiv:2102.00106v36 March 2021.

[88] Giaquinta, M., *Multiple Integrals in the Calculus of Variations and Nonlinear Elliptic Systems*, Annals of Mathematics Studies **105**, Princeton University Press, Princeton, 1983.

[89] Gilbarg, D. and Trudinger, N. S., *Elliptic Partial Differential Equations of Second Order*, Springer-Verlag, Berlin-Heidelberg-New York, 1977.

[90] Grisvard, P., *Elliptic Problems in Nonsmooth Domains*, Monographs and Studies in Mathematics **24**, Pitman, Boston, 1985.

[91] Guzu, D., Kapitanski, L. and Laptev, A., Hardy inequalities for discrete magnetic Dirichlet forms, Pure Appl. Funct. Anal., to appear.

[92] Hadwiger, H., *Vorlesungen über Inhalt, Oberfläche und Isoperimetrie*, Springer, Berlin/Gòttingen/Heidelberg, 1957.

[93] Hajłasz, P., Pointwise Hardy inequalities, Proc. Amer. Math. Soc. **127**(2) 1999, 417–423.

[94] Hardy, G. H., An inequality between integrals, Messinger Math. **54** (1925), 150–156.

[95] Haroske, D. D. and Triebel, H., *Distributions, Sobolev Spaces, Elliptic Equations*, European Mathematical Society, Zürich, 2008.

[96] Herbst, I., Spectral theory of the operator $\left(p^2 + m^2\right)^{1/2} - ze^2/r$, Comm. Math. Phys. **53**(3) (1977), 285–294.

[97] Hewitt, E. and Stromberg, K., *Real and Abstract Analysis*, Springer, New York, 1965.

[98] Hoffmann-Ostenhof, M., Hoffmann-Ostenhof, T. and Laptev, T., A geometrical version of Hardy's inequality, J. Funct. Anal. **189** (2002), 539–548.

[99] Hurri-Syränen, R. and Vähäkangas, A. V., On fractional Poincaré inequalities, J. Anal. Math. **120** (2013), 85–104.

[100] Hurri-Syränen, R. and Vähäkangas, A. V., Fractional Sobolev-Poincaré and fractional Hardy inequalities in unbounded John domains, Mathematika **61**(2) (2015), 385–401.

[101] Hurri-Syränen, R., Weighted fractional Poincaré type inequalities, Colloq. Math. **157** (2019).

[102] Iannizzotto, A., Mosconi, S. and Squassina, M., Global Hölder regularity for the fractional p-Laplacian, Rev. Mat. Iberoam. **32** (2016), 1353–1392.

[103] Iannizzotto, A. and Squassina, M., Weyl-type laws for fractional p−eigenvalue problems, Asymptotic analysis **88** (2014), 233–245.

[104] Itoh, J.-I. and Tanaka, M., The Lipschitz continuity of the Lipschitz function to the cut locus, Trans. Amer. Math. Soc. **353**(1) (2001), 21–40.

[105] Iwaniec, T. and Martin, G., Riesz transforms and related singular integrals, J. Reine Angew. Math. **473** (1996), 25–57.

[106] Janson, S., Taibleson, M. and Weiss, G., Elementary characterizations of the Morrey-Campanato spaces. In: *Harmonic Analysis* (Cortona 1092), Lecture Notes in Math., vol 1992, pp. 101–114, Springer, Berlin (1983).

[107] Karadzhov, G. E., Milman, M. and Xiao, J., Limits of higher-order Besov spaces and sharp reiteration theorems, J. Functional Anal. **221** (2005), 323–339.

[108] Kinnunen, J., Lehrbäck, J. and Vähäkangas, A. V., *Limits of Maximal Function Methods for Sobolev Spaces*, AMS Math. Surveys, American Math. Soc., Providence, RI, 2021.

[109] Kolyada, V. and Lerner, A., On limiting embeddings of Besov spaces, Studia Math. **171**(1) (2005), 1–13.

[110] Kufner, Maligranda, L. and Persson, L.-E., *The Hardy Inequality: About its History and Some Related Results*, Vydavatelský servis, Pilsen, 2007.

[111] Kuusi, T., Mingione, G. and Sire, Y., Nonlocal equations with measure data, Comm. Math. Phys. **337**(3) (2015), 1317–1368.

[112] Kwaśnicki, M., Ten equivalent definitions of the fractional Laplace operator, Fract. Calc. Appl. Anal. **20** (2017), 51–57.

[113] Lamberti, P. D. and Pinchover, Y., L^p Hardy inequality on $C^{1,\gamma}$ domains, Annali Scuola Normale Superiore Pisa **19** (2019), 1135–1159.

[114] Landau, E., A note on a theorem concerning positive terms, J. Lond. Math. Soc. **1** (1926), 38–39.

[115] Laptev, A. and Sobolev, A. V., Hardy inequalities for simply connected planar domains, Amer. Math. Soc. Transl. Ser. 2 **225** (2008), 133–140.

[116] Laptev, A. and Weidl, T., Hardy inequalities for magnetic Dirichlet forms, Oper. Theory: Adv. Appl. **108** (1999), 299–305.

[117] Lang, J. and Musil, V., Strict s–numbers of non-compact Sobolev embeddings into continuous functions, Constr. Approx. **50**(2) (2019), 271–291.

[118] Lang, J. and Nekvinda, A., Embeddings between Lorentz sequence spaces are strictly singular, arXiv:2104.00471v1 (April 2021).

[119] Lefèvre, P. and Rodríguez-Piazza, L., Finitely strictly singular operators in harmonic analysis and function theory, Adv. Math **255** (2014), 119–152.

[120] Lehrbäck, J., Pointwise Hardy inequalities and uniformly fat sets, Proc. Amer. Math. Soc. **1.36**(6) (2008), 2193–2200.

[121] Lehrbäck, J., Weighted Hardy inequalities and the size of the boundary, Manuscripta Math. **127** (2008), 249–273.

[122] Lewis, J. L., Uniformly fat sets, Trans. Amer. Math. Soc. **308** (1988), 177–196.

[123] Lewis, R. T., Singular elliptic operators of second order with purely discrete spectra, Trans. Amer. Math. Soc. **271** (1982), 653–666.

[124] Lewis, R. T., Li, J. and Li, Y., A geometric characterization of a sharp Hardy inequality, J. Funct. Anal. **262**(7) (2012), 3159–3185.

[125] Lieb, E. H., On the lowest eigenvalue of the Laplacian for the intersection of two domains, Invent. Math. **74**(3) (1983), 441–448.

[126] Lieb, E. H. and Loss, M., *Analysis*, 2nd edition, Amer. Math. Soc. Graduate Studies in Math. **14**, 2001.

[127] Lindgren, E. and Lindqvist, P., Fractional eigenvalues, Calc. Var. Partial Differential Equations **49** (2014), 795–826.

[128] Lindqvist, P., Notes on the *p*-Laplace equation, Report. University of Jyväskylä Department of Mathematics and Statistics, 2006.

[129] Li, Y., and Nirenberg, L., The distance to the boundary, Finsler geometry and the singular set of viscosity solutions of some Hamilton-Jacobi equations, Comm. Pure Appl. Math. **18**(1) (2005), 85–146.

[130] Loss, M. and Sloane, C., Hardy inequalities for fractional inequalities on general domains, J. Functional Anal. **259** (2010), 1369–1379.

[131] Mantegazza, C. and Mennici, A. C., Hamilton-Jacobi equations and distance functions on Riemannian manifilds, Appl. Math. Optim. **47** (2003), 1–25.

[132] Marcus, M., Mizel, V., and Pinchover, Y., On the best constant in Hardy's inequality in $\mathbb{R}^n$, Trans. Amer. Math. Soc. **350**(8) (1998), 3237–3255.

[133] Matskewitch, T. and Sobolevskii, P., The best possible constant in generalized Hardy's inequality for convex domains in $\mathbb{R}^n$, Nonlinear Anal. Theory Methods Appl. **28**(9) (1997), 1601–1610.

[134] Maz'ya, V. and Shaposhnikova, T., On the Bourgain, Brezis and Mironescu theorem concerning limiting embeddings of fractional Sobolev spaces, J. Functional Anal. **195** (2002), 230–238; corrig. ibid. **201** (2003), 298–300.

[135] Maz'ya, V., *Sobolev Spaces*, Springer, Berlin, Heidelberg, 1985.

[136] Milman, M., Notes on limits of Sobolev spaces and the continuity of interpolation scales, Trans. Amer. Math. Soc. **357**(9) (2005), 3425–3442.

[137] Mosconi, S. and Squassina, M., Recent progresses in the theory of nonlinear nonlocal problems, Bruno Pini Math. Analysis Sem. **7** (2016), 147–164.

[138] Moser, J., On Harnak's theorem for elliptic differential equations, Comm. Pure Appl. Math. **16** (1961), 577–591.

[139] Motzkin, T. S., Sur quelques propriétés charactéristiques des ensembles convex, Atti Real. Accad. Naz. Lindcei Rend. Cl. Sci. Fis. Mat. Natur. Serie VI **21** (1935), 562–567.

[140] Naibullin, R., Hardy and Rellich type inequalities with remainders, to appear.

[141] Nečas, J., *Les méthodes directes en théorie des équations elliptiques*, Masson, Paris, 1967.

[142] Nezza, E. di, Palatucci, G. and Valdinoci, E., Hitchhiker's guide to the fractional Sobolev spaces, Bull. Sci. Math. **136**(5) (2012), 521–573.

[143] Owen, M. P., The Hardy–Rellich inequality for polyharmonic operators, Proc. Roy. Soc. Edin. **129 A** (1999), 825–839.

[144] Perera, K., Agarwal, R. P. and O'Regan, D., *Morse theoretic aspects of p-Laplacian type operators*, Mathematical surveys and monographs, Amer. Math. Soc., Providence, RI, 2010.

[145] Peetre, J., Espaces d'interpolation et théorème de Soboleff, Ann. Inst. Fourier **16** (1966), 279–317.

[146] Pick, L., Kufner, A., John, O. and Fučík, S., *Function Spaces*, Vol. 1 (2nd revised and extended edition), De Gruyter, Berlin/Boston, 2013.

[147] Pietsch, A., *History of Banach Spaces and Linear Operators*, Birkhäuser, Boston, 2007.

[148] Pinchover, Y. and Goel, D., On weighted L^p-Hardy inequalities on domains in $\mathbb{R}^n$, arXiv: 2012.12860v2.

[149] Prats, M. and Saksman, E., A $T(1)$ theorem for fractional Sobolev spaces on domains, J. Geom. Anal. **27**(3) (2017), 2490–2518.

[150] Ponce, A., A new approach to Sobolev spaces and connections to Γ-convergence, Calc. Var. Partial Diff. Eqs. **19** (2004), 229–255.

[151] Qui, H. and Xiang, M., Existence of solutions for fractional p-Laplacian problems via Leray-Schauder nonlinear alternative, Boundary value problems (2016), doi 10.1186/s13661-016-0593-8.

[152] Rafeiro, H., Samko, N. and Samko, S., Morrey-Campanato spaces: an overview, Operator Theory, Advances and Applications **228** (2013), 293–323.

[153] Rellich, F., Halbeschränkte Differential operatoren höherer Ordnung. In *Proceedings of the International Congress of Mathematicians 1954*, vol.III, pp. 243–250, Noordhoff, Groningen (1956).

[154] Robinson, D. W., Hardy and Rellich inequalities on the complement of convex sets, J. Austr. Math. Soc. **108**(1) (2020), 98–119.

[155] Robinson, D. W., The weighted Hardy constant, arXiv:2103.07.848v1 14 Mar. 2021.

[156] Samko, S., Best constant in the weighted Hardy inequality: the spacial and spherical version, Fractional Calculus and Applied Analysis **8**(1) (2005), 39–52.

[157] Schilling, R. L., *Measure, Integrals and Martingales*, Cambridge University Press, Cambridge, 2005.

[158] Schilling, R. and Kühn, F., *Counterexamples in Measure and Integration*, Cambridge University Press, Cambridge, 2021.

[159] Slavíková, L., Almost compact embeddings, Math. Nachr. **285** (2012), 1500–1516.

[160] Sloane, C. A., A fractional Hardy–Sobolev–Maz'ya inequality in the half-space, Proc. Amer. Math. Soc. **139** (2011), 4003–4016.

[161] Slobodeckij, L. N., Generalised Sobolev spaces and their applications to boundary value problems of partial differential equations (Russian), Leningrad Gos. Ped. Inst. Učep. Zap. **197** (1958), 54–112.

[162] Sobolevskii, P. E., Hardy's inequality for the Stokes problem, Nonlinear Analysis, Methods & Applications **30**(1) (1997), 129–145.

[163] Solomyak, M. Z., A remark on the Hardy inequalities, Integr. Equ. Oper. Theory **19** (1994), 120–124.

[164] Stromberg, K. R., *An Introduction to Classical Real Analysis,* Wadsworth, Belmont, 1981.

[165] Talenti, G., An inequality between $u*$ and $|\nabla u *|$, Inst. Series Num. Math., **103** (1992), 175–182.

[166] Thomas, J. C., *Some problems associated with sum and integral inequalities*, Ph.D. thesis, Cardiff University, Wales (2007).

[167] Tidblom, J., A geometrical version of Hardy's inequality for $W_0^{1,p}(\Omega)$, Proc. Amer. Math. Soc. **132**(8) (2004), 2265–2271.

[168] Tidblom, J., A Hardy inequality in the half-space, Research report in mathematics 3, Department of Mathematics, University of Stockholm, 2004.

[169] Triebel, H., *Theory of Function Spaces*, Birkhäuser, Basel, 1983.

[170] Triebel, H., *Interpolation Theory, Function Spaces, Differential Operators* (2nd edition), Barth, Heidelberg, 1995.

[171] Triebel, H., *The Structure of Functions*, Birkhäuser, Basel, 2001.
[172] Yafaev, D., Sharp constants in the Hardy–Rellich inequalities, J. Functional Anal. **168**(1) (1999), 121–144.
[173] Zhou, Y., Fractional Sobolev extension and imbedding, Trans. Amer. Math. Soc. **367**(2) (2015), 959–979.

Symbol Index

Author Index

Subject Index

Printed in the United States
by Baker & Taylor Publisher Services